CRITICAL EVALUATION OF SOME EQUILIBRIUM CONSTANTS INVOLVING ORGANOPHOSPHORUS EXTRACTANTS

INTERNATIONAL UNION OF PURE AND APPLIED CHEMISTRY
ANALYTICAL CHEMISTRY DIVISION
COMMISSION ON EQUILIBRIUM DATA

CRITICAL EVALUATION OF SOME EQUILIBRIUM CONSTANTS INVOLVING ORGANOPHOSPHORUS EXTRACTANTS

Y. MARCUS

BUTTERWORTHS

THE BUTTERWORTH GROUP

ENGLAND
Butterworth & Co (Publishers) Ltd
London: 88 Kingsway, WC2B 6AB

AUSTRALIA
Butterworths Pty Ltd
Sydney: 586 Pacific Highway, NSW 2067
Melbourne: 343 Little Collins Street, 3000
Brisbane: 240 Queen Street, 4000

CANADA
Butterworth & Co (Canada) Ltd
Toronto: 2265 Midland Avenue, Scarborough, M1P 4S1

NEW ZEALAND
Butterworths of New Zealand Ltd
Wellington: 26–28 Waring Taylor Street, 1

SOUTH AFRICA
Butterworth & Co (South Africa) (Pty) Ltd
Durban: 152–154 Gale Street

First published 1974

ISBN 0 408 70646 5

Printed in England by Page Bros (Norwich) Ltd, Norwich

Critical Evaluation of Some Equilibrium Constants Involving Organophosphorus Extractants*
(Report to IUPAC Commission V.6†)

CONTENTS

* The work was supported by the United States National Bureau of Standards as a part of its National Standard Reference Data series.

† *Titular Members (1973):* D. N. Hume (Chairman); F. J. C. Rossotti (Secretary); Members: M. T. Beck, E. Högfeldt, A. S. Kertes, G. N. Nancollas, D. D. Perrin, J. Stary.
Associate Members: S. Ahrland, G. Anderegg, S. Bruckenstein, H. Freiser, D. L. Leussing, L. B. Rogers, K. Yamasaki, Yu. A. Zolotov.
National Representatives: Y. Marcus (Israel), K. B. Yatzimirskii (USSR), A. E. Martell (USA).

Introduction

A critical evaluation, supplementary to the compilation "Equilibrium Constants of Liquid-Liquid Distribution Reactions (Introduction, and Part I: Organophosphorus Extractants)," should enhance the usefulness of this compilation. It turned out that for the great majority of distribution systems the compiled constants, as reported in the original publications, must serve, and only for a minority of them has it been possible to embark on a critical evaluation. This is due to the great variety of reactions that must be considered, and to the fact that a given reaction has often been examined only by one author, the results not having been confirmed, nor criticized by others with comparable experimental data. A number of reactions, however, have been fairly well documented, and these can be critically discussed.

The reactions evaluated comprise those from List 1 for which data are available; see Table of Contents. The reactions

$$H^+ + NO_3^- + \overline{TB'PO} = \overline{HNO_3 \cdot TB'PO} ,$$

$$UO_2^{2+} + 2NO_3^- + 2\ \overline{TO'PO} = \overline{UO_2(NO_3)_2 \cdot (TO'PO)_2}$$

and

$$M^{3+} + 3NO_3^- + 3\ \overline{TBP} = \overline{M(NO_3)_3(TBP)_3} \quad (M = \text{rare earth})$$

have also been closely examined. Because of lack of data, no critical evaluation could be made.

List 1: Reactions

$S = \overline{S}$, the solvent distribution reaction

$\overline{S} + H_2O = \overline{S \cdot H_2O}$, and occasionally higher hydrate formation

$2\overline{S} = \overline{S}_2$, and occasionally higher aggregation

$\overline{HS} + \overline{S'} = \overline{HS \cdot S'}$, adduct formation

$H^+ + S^- = HS$, the solvent association for acidic solvents

$H^+ + NO_3^- + \overline{S} = \overline{HNO_3 \cdot S}$

$H^+ + Cl^- + \overline{S} = \overline{HCl \cdot S}$

$UO_2^{2+} + 2NO_3^- + 2\overline{S} = \overline{UO_2(NO_3)_2S_2}$

$M^{m+} + mA^- + n\overline{S} = \overline{MA_mS_n}$

$M^{m+} + m\,\overline{H_2S_2} = \overline{M(HS_2)_m} + mH^+$

Here, M^{m+} is a metal ion, A^- a univalent ligand and S, S' and HS are solvents. The solvents are taken from List 2, again, as far as data are available. Abbreviations of solvents and diluents follow Marcus et al., J. Inorg. Nucl. Chem., 33, 863 (1971).

List 2: Solvents

Tri-n-butyl phosphate , TBP

Dibutyl butylphosphonate , DBB'P

Tri-n-butyl phosphine oxide , TB'PO

Tri-n-octyl phosphine oxide , TO'PO

Di-n-butyl phosphoric acid , (H)DBP

Bis (2-ethylhexyl) phosphoric acid , (H)DE-HP

The symbols used in the discussion of the equilibrium constants are given in List 3, and the few additional symbols used are defined in the text where encountered.

List 3: Symbols

a_A	activity of A, rational scale unless otherwise noted
C_A	total concentration of A, molar scale
D_A	distribution ratio of A, total molar concentration of A in the organic phase divided by its total molar concentration in the aqueous phase
d	density
f_A	activity coefficient of A, rational (mole fraction) scale
$\Delta G^\circ_{(i)}$	standard Gibbs energy change for one gram-formula of reaction (i)
$\Delta H^\circ_{(i)}$	standard enthalpy change for one gram-formula of reaction (i)
$K_{(i)}$	thermodynamic equilibrium constant of reaction (i) (molar scale understood, other scales noted)
M_A	gram-molecular (formula) weight of A
m_A	molality of A
$Q_{(i)}$	equilibrium concentration quotient of reaction (i)
R	the gas constant, 8.314 JK^{-1} (1.987 cal K^{-1}) mol^{-1}
$\Delta S^\circ_{(i)}$	standard entropy change for one gram-formula of reaction (i)
T	temperature in K
V_A	molar volume of A
X_A	mole fraction of A
y_A	activity coefficient of A, molar scale
β_i	overall (mononuclear) complex formation constant for the addition of i ligands
γ_A	activity coefficient of A, molal scale
μ	ionic strength
$^\circ$	standard state of pure component
∞	reference state of infinite dilution of the solute(s)

— a species in, or a quantity referring to, the organic phase

(A) thermodynamic activity of species A (molar scale understood, other scales noted)

[A] concentration of species A, molar scale

The equilibrium constants have almost invariably been reported as equilibrium quotients, measured and averaged over a more or less wide range of the concentrations of the reactants involved. The equilibrium quotients have usually been calculated from the dependence of the distribution ratio of one reactant between an organic and an aqueous phase on the concentration of either itself, or of one or more other reactants in one of the phases. These overall concentrations are usually made up by volume, using 99%+ pure reagents, and concentrations, in the macro range, are usually determined by titrimetry, so that a precision (i.e. standard deviation) of ± 0.5% for a single measurement is attained by very careful work, while ±1.0% would be a more generally applicable figure. At low concentrations, say 10^{-3} M or lower, a spectrophotometric, or more usually a radiometric analytical determination is used. These have lower precision, ±2% would be a common measure. Distribution ratios depend on the ratio of two such figures, and often (for $D > 10^3$ or $D < 10^{-3}$) one of these represents a very low concentration, known to only ±10% or perhaps worse. The general precision of distribution coefficients is therefore ±5% or worse, and these usually determine the overall precision of the equilibrium quotient. Since the latter depends usually on 10 or more experimental points of a distribution curve, the author may perform a least-squares error calculation, and come up with a standard deviation estimate, which at best is as small as the precision index of a single point, i.e. ±5% or 0,02 units in log Q. Few reports attain this precision.

Many of the equilibrium quotients reported in the literature are based on data of poorer precision, and the individual points agree less well with a smooth distribution curve, so that a least-squares analysis would indicate a standard deviation of, say ± 0.06 units in log Q.

Systematic errors are detected by noting a trend in the equilibrium quotient over the concentration range studied. A comparison of data from several sources, over as wide a concentration range as practical, helps to eliminate such errors, which are due mainly to the assumption that a given reaction, yielding a given species, uniquely or even predominantly represents the equilibrium state under the given conditions.

When the equilibrium quotient of an author is reported in this review, his estimate of the standard deviation is checked for its compatibility with the quality of his experimental methods and the data reported. When the data of several authors are compared, the constants are averaged, weighted inversely proportional to their standard deviation, and the error quoted is the standard least-squares deviation of this average or the least of the standard deviations of the reported constants, whichever is greater.

In a table at the end of the discussion of each reaction, each solvent is given the finally-evaluated best constant (or constants, in the case of different diluents). The reaction and the standard state to which the constant pertains are given as sub- and superscripts. The reliability of the constant is indicated by the standard deviation assigned to it according to the procedure discussed above. Completely unreliable constants will not appear in the tables, but occasionally, for the purpose of comparison with other constants (e.g. for other diluents), some less reliable values are entered and marked thus: (?). These need be considered with great care, and cannot be taken as being endorsed by the present reviewer.

I. Solvent: Di-n-butylphosphoric acid ((H)DBP), HS

Reaction: $H^+ + S^- = HS$ (in aqueous solutions) (I-1)

(H)DBP is fairly well water-soluble, 0.0818±0.0002 M at 25°C (57D, 59H, 66M), so that its acid dissociation, the reverse of reaction (I-1), can be easily studied. Both the classical pH - titration method and a radiometric (or other) distribution method have been applied.

The equilibrium constant of reaction (I-1) is

$$K_{(1)} = (HS)/(H^+)(S^-) = Q_{(1)} y_{HS}/y_{\pm HS}^2 \qquad (I-2)$$

where the right hand side of (I-2) defines the equilibrium quotient $Q_{(1)}$. With a constant ionic medium the activity coefficient ratio is constant, and so is therefore $Q_{(1)}$. An early determination (43K), $\log Q_{(1)} = 1.72$ does not agree with all the later work, and although the method used seems to have been valid, yielding a reasonable value for the corresponding quotient for $H_2PO_4^- + H^+ = H_3PO_4$, $\log Q = 1.97$, the barium dibutylphosphate used must have been contaminated with a less strongly acidic salt, such as the monobutylphosphate, or even excess sulfuric acid. Some other values of $Q_{(1)}$ were obtained without a constant ionic medium: $\log Q_{(1)} = 1.036$ for saturated aqueous HS with no additional electrolyte (57D), $\log Q_{(1)} = 1.21 \pm 0.06$ for dilute (10^{-3} to 10^{-2} M) solutions of HS in water (66K), and $\log Q_{(1)} = -1.29 \pm 0.04$ (66M) for 0.1 to 0.8 M HCl, and will not be considered in the following.

At concentrations of $C_{HS} > 5\times10^{-3}$ M, there is a complication from the formation of the dimer H_2S_2 and the dimer-ion HS_2^- (57D, 63A, 67K, 66M, 68K), but below $C_{HS} = 10^{-3}$ M (corresponding to $\overline{C}_{HS}$ from 10^{-5} M to 10^{-4} M in typical distribution experiments with, say, carbon tetrachloride at pH 2 to 0) this complication should be absent. When the ionic medium is 0.1 M, similar results were obtained with nitrate, perchlorate and bisulfate (57D, 65U, 70U), giving the value shown in Table I-1, which should be reliable.

In 1 M media the results are discordant (57D, 59H, 67K, 68K). The differences are not between different ionic media - nitrate and perchlorate - as might have been expected, so much as between results obtained from distribution measurements with different organic solvents (68K) and at different C_{HS} (67K). There is no explanation for the different log $Q_{(1)}$, ranging from 1.04±0.06 to 1.20±0.07 reported for differing C_{HS}, all below 10^{-3} M, and already corrected for dimeric H_2S_2 in the aqueous phase (with an unreasonably high dimerization constant), except for the species HS_2^- (63A, 66M) having been neglected. The curves log D_{HS} vs. pH should be parallel at a given $\overline{C}_{HS}$ (also at a given initial $\overline{C}_{HS}$, provided only a small fraction goes into the aqueous phase, or at such low $\overline{C}_{HS}$ that only monomers are important), and their observed (68K) non-parallel course, leading to different estimates of log $Q_{(1)}$ in 1 M medium cannot be explained. From the considerations below (see also Fig. I-1), the estimate 1.00 (57D, 59H) seems to be nearer the true value of log $Q_{(1)}$ (1.0 M medium) than the higher (67K) or lower (68K) estimates.

Attempts have been made to estimate, partly, the activity coefficient quotient in (I-2) (57D, 67K). The following may lead to as good an estimate of $K_{(1)}$, at zero ionic strength, as any other. The extended Debye-Hückel formula for $y_{\pm HS}$, as suggested by Vasilev (62V), and the Setchenov formula for y_{HS} give

$$\log K_{(1)} = \log Q_{(1)}(\mu) + 1.018\sqrt{\mu}/(1 + 1.6\sqrt{\mu}) + (b' - 2b)\mu \qquad (I\text{-}3)$$

The data shown in Fig. I-1 are consistent with the value 0.20 for the parameter (b'-2b) and the extrapolated value of log $K_{(1)}$ shown in Table I-1. This value is also in good agreement with the value log $K_{(1)}$ = 1.17 estimated from single-ion activity coefficients (57D). It must however be commented that the data at higher ionic strengths (63A), used mainly to fix the value of the parameter (b' - 2b), are not very reliable, since for each ionic strength only four experimental data points were given, from which no less than five equilibrium constants (for dissociation, dimerization and distribution equilibria) have been calculated. The estimates of the activity of HS in TBP solutions that have been used to calculate $K_{(1)}$ values in 0.1 to 0.8 M HCl (66M) have been shown to be unreliable (71K).

Table I-1

The Acid Association Constants of (H)DBP in Aqueous Solution at 25°

$\log Q_{(1)}$ (0.1 M medium)	=	1.00±0.02
$\log Q_{(1)}$ (1.0 M medium)	=	1.0
$\log K_{(1)}$ (pure water)	=	1.19±0.03

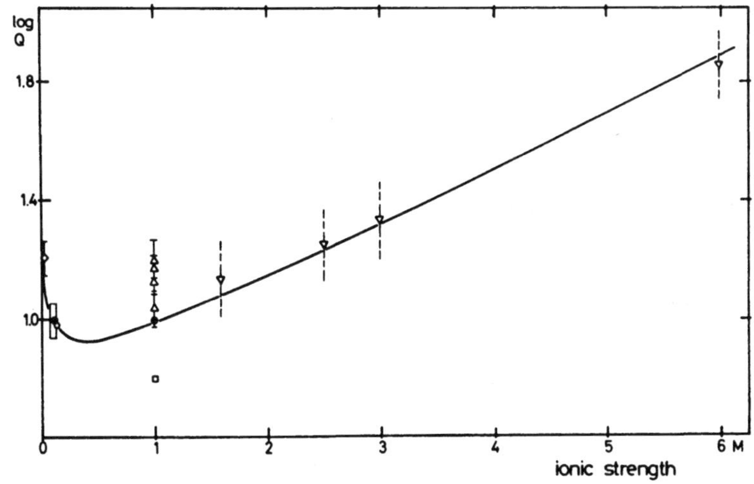

Acid association quotients of HS, log $Q_{(1)}$, as a function of ionic strength. Continuous line calculated from eq. (I-3) with log $K_{(1)}$ = 1.19 and (b' - 2b) = 0.20. The data points show their estimated standard deviations: ● 57Da , 59H; o 68L; □ 68Kb (all these are ±0.01); ▯ 65U, 70U; Δ 67K; ∇ 63A: ◇ 66Ka

References:

43K Kumler, W.D. and Eiler, J.J., J. Am. Chem. Soc., 65, 2355 (1943).

51D Dyrssen, D., Acta Chem. Scand., 11, 1771 (1957).

59H Hardy, C.J. and Scargill, D., J. Inorg. Nucl. Chem., 11, 128 (1959).

62V Vasilev, V.P., Zh. Neorg. Khim., 7, 1788 (1962).

63A Ampelogova, N.I., Radiokhimiya, 5, 562 (1963).

65U Ul'yanov, V.S. and Sviridova, R.A., Radiokhimiya, 7, 538 (1965).

66K Kiss, A. and Hegedus, A.J., Mikrochim. Acta, 1966 (4-5), 771.

66M Mikhailov, V.A., Grigoreva, E.F. and Kharchenko, S.K., Zh. Fiz. Khim., 40, 1638 (1966).

67K Kuča, L., Coll. Czech. Chem. Comm., 32, 729 (1967).

68K Kolařik, Z., Hejna, J. and Pankova, H., J. Inorg. Nucl. Chem., 30, 2795 (1968).

68L Liem, D.H., Acta Chem. Scand., 22, 753 (1968).

70U Ul'yanov, V.S. and Sviridova, R.A., Radiokhimiya, 12, 47 (1970).

71K Kolařik, Z., Solv. Extr. Rev., 1, 1 (1971).

II. Solvent: Di-n-butylphosphoric acid ((H)DBP), HS

Reaction: $HS = \overline{HS}$ (II-1)

The distribution reaction of HS has usually been studied in parallel with its dimerization reaction in organic solvents, so that the information that has been published is extensive. The distribution method, often with radioactively-tagged (^{32}P)HS, has been commonly used, and the side reactions of dissociation to H^+ and S^- in the aqueous phase and dimerization to $\overline{H_2S_2}$ in the organic phase have been taken into account. However, some complicating side reactions, occuring at high concentrations of HS, such as formation of H_2S_2 and HS_2^- in the aqueous phase, or $\overline{H_nS_n}$ ($n > 2$) in the organic phase, have not always been corrected for where necessary, leading to errors. Further uncertainty is introduced by the failure of authors to standardize on one constant ionic medium, so that the salting-out properties of the aqueous phase cannot always be definitely and accurately taken into account.

The salting-out behavior has been studied in HNO_3-$NaNO_3$ mixtures (67K) for CCl_4 as solvent and for $HClO_4$-$NaClO_4$ mixtures (57D) for methyl-isobutyl ketone as solvent. For several other solvents, different authors have obtained the equilibrium quotient for reaction (II-1) at 0.1 M and 1.0 M for both nitrate and perchlorate media, so that

$$\Delta\log Q_{(1)} = \log Q_{(1)} \text{ (1.0 M medium)} - \log Q_{(1)} \text{(0.1 M medium)} \quad \text{(II-2)}$$

can be computed. It turns out to be independent of the organic solvent as expected, and also of the chemical nature of the medium, whether nitrate or perchlorate. The mean value is

$$\Delta \log Q_{(1)} = 0.14 \pm 0.06 \quad \text{(II-3)}$$

Data obtained in a 1.0 M medium have been normalized to a 0.1 M medium therefore by subtracting 0.14 from log Q (1.0 M medium), and in the following, data will be referred to a 0.1 M ionic medium as the standard state of the aqueous phase. The equilibrium constant of reaction (II-1) is

$$K_{(1)} = (\overline{HS})/(HS) = [\overline{HS}]\overline{y}_{HS}/[HS]y_{HS} = Q_{(1)}\overline{y}_{HS}/y_{HS} = Q'_{(1)}/y_{HS} \qquad \text{(II-4)}$$

Since y_{HS} is taken to be a constant in the standard medium, $Q'_{(1)} = Q_{(1)}\overline{y}_{HS}$ is seen to be a constant, but $Q_{(1)}$ should vary inversely with $\overline{y}_{HS}$. Even if all complicating reactions have been taken into account, $Q_{(1)}$ should thus still be a function of $\overline{C}_{HS}$, and only approach a constant value

$$Q^{\infty}_{(1)} = \lim\ (\overline{C}_{HS} \rightarrow 0)\ Q_{(1)} \quad \text{(0.1 M medium)} \qquad \text{(II-5)}$$

as the standard reference state, the wet organic solvent in equilibrium with the 0.1 M ionic medium, is approached.

As a basis for the selection of the best $Q^{\infty}_{(1)}$ values for the various solvents, their conformity to a correlation with the solubility parameter of the solvents (71K) has been used. The curve which expresses the data best obeys the following expression

$$\log Q^{\infty}_{(1)} = -2.46 \pm 0.12 + (0.02 \pm 0.03)(\delta - 7.00) + (0.43 \pm 0.02)(\delta - 7.00)^2 \qquad \text{(II-6)}$$

and the data that conform to this relationship have been weighted more than those that do not. The resulting selected values are shown in Table II-1.

Solvents which have donor oxygen atoms and those which are polar and have high δ values (chlorobenzene, nitrobenzene and bromoform) do not conform to expression (II-6). Since the solubility parameter is only one of several parameters that characterize the diluents, conformity to expression (II-6) should not be used as a strict criterion in judging the values.

A reaction that is not considered in the above is the "solvation" of HS with the diluent, which would be of particular importance with those solvents containing donor oxygen atoms. A case has been argued for considering $\overline{HS}$ to react with the diluent S' according to

$$\overline{HS} + \overline{S'} = \overline{HS \cdot S'} \qquad (II-7)$$

so that (60D)

$$\overline{C}_{HS} = [\overline{HS}][1 + K_{(7)}[\overline{S'}](\overline{y}_{HS}\overline{y}_{S'}/\overline{y}_{HS \cdot S'})] \qquad (II-8)$$

As an approximation, $[\overline{S'}]$ was set equal to $\overline{C}_{S'}$, the molar concentration of the diluent, or at low $\overline{C}_{HS}$ (and neglecting extracted water)

$$[\overline{S'}] \approx 1000\ d^{o}_{S'}/M_{S'}\ , \qquad (II-9)$$

where $d^{o}_{S'}$ is the density of pure diluent and $M_{S'}$ its formula weight. Also

$$K_{(7)}(\overline{y}_{HS}\overline{y}_{S'}/\overline{y}_{HS \cdot S'}) = Q_{(7)} \qquad (II-10)$$

was considered a constant, while

$$Q^{o}_{(1)} = [\overline{HS}]^{o}/[HS] \qquad (II-11)$$

was also considered a constant (Nernst's law), independent of the solvent, leading at the limit $\overline{C}_{HS} \rightarrow 0$ to

$$\log Q^{\infty}_{(1)} = \log Q^{o}_{(1)} + \log (1 + Q_{(7)}\overline{C}_{S'}) \qquad (II-12)$$

Since, however, $Q^{o}_{(1)}$ need not be a constant independent of the solvent,* there is no valid way to express the log $Q^{\infty}_{(1)}$ values (Table II-1) in terms of separate log $Q^{o}_{(1)}$ and log $Q_{(7)}$ values. Indeed, the occurence of reaction (II-7), with

*Indeed, for inert solvents for which $Q_{(7)} \approx 0$ is strongly indicated, log $Q^{o}_{(1)}$ is seen by conforming to eq. (II-6) to depend on the solvent.

a 1:1 adduct in the undiluted solvent S', has not been proven, or has perhaps been counterindicated. Therefore, efforts to refine eqs. (II-8) and (II-12) will probably be wasted.

Table II-1

Distribution Equilibrium Quotients for (H)DBP between 0.1 M Aqueous Media and Organic Diluents at 25°

Diluent	$\log Q^{\infty}_{(1)}$	Refs.
iso-Octane	-2.53±0.05	70U
n-Hexane	-2.36±0.02	60Da,70U
n-Octane	-2.37±0.04	65U,70U
Kerosene	-2.10±0.07	59H
cyclo-Hexane	-1.85±0.07	68K
Carbon tetrachloride	-1.33±0.05	66L,67K,68L
Toluene	-0.79±0.07	61G,68K
o-Xylene	-0.80±0.07	68K
Benzene	-0.23±0.05	68K
Chloroform	+0.26±0.04	57D,59H,60Da,70U
Nitrobenzene	-0.14±0.07 (?)	59H,68K
Chlorobenzene	-0.69±0.07 (?)	68K
Bromoform	-0.28±0.07 (?)	68K
Dibutyl ether	-0.15±0.03	59D,60D
Di-iso-propyl ether	+0.52±0.03	60Da
Methyl isobutyl ketone	+0.37±0.03	57D,68K
Tri-n-butylphosphate	+1.97±0.03	60D
Methyl isobutyl carbinol	+2.21±0.05	60D

References:

57D Dyrssen, D., Acta Chem. Scand., 11, 1771 (1957).

59D Dreze, Ph., Bull. Soc. Chim. Belges, 68, 674 (1959).

60D Duyckaerts, G., Dreze, Ph. and Simon, A., J. Inorg. Nucl. Chem., 13, 332 (1960).

60Da Dyrssen, D. and Liem, D.H., Acta Chem. Scand., 14, 1091 (1960).

61G Greenfield, B.F. and Hardy, C.J., J. Inorg. Nucl. Chem., 21, 359 (1961).

65U Ul'yanov, V.S. and Sviridova, R.A., Radiokhimiya, 7, 538 (1965).

66L Liem, D.H. and Dyrssen, D., Acta Chem. Scand., 20, 272 (1966).

67K Kuča, L., Coll. Czeck. Chem. Comm., 32, 729 (1967).

68K Kolařik, Z., Hejna, J. and Pankova, H., J. Inorg. Nucl. Chem., 30, 2795 (1968).

68L Liem, D.H., Acta Chem. Scand., 22, 753 (1968).

70U Ul'yanov, V.S. and Sviridova, R.A., Radiokhimiya, 12, 47 (1970).

71U Kolařik, Z., Solv. Extra. Rev., 1, 1 (1971).

III. Solvent: Di-n-butylphosphoric acid ((H)DBP) , HS

Reaction: $$2\ \overline{HS} = \overline{H_2S_2} \qquad \text{(III-1)}$$

The dimerization of HS in organic solvents is a well documented reaction. The more "inert" the diluent, the more pronounced is the tendency of HS to react with itself, leading in paraffinic hydrocarbons at sufficiently high concentrations ($\overline{C}_{HS} > 0.05$ M) to higher oligomers (60D), such as a tetramer in hexane (68L) or a hexamer in cyclohexane (68K). Conversely, the more "active" the diluent, the more it can react with HS monomers through dipole interactions or hydrogen bonding, thus diminishing the tendency towards self-interaction, because of the tremendous concentration advantage of the diluent.

The equilibrium constant of reaction (III-1) is

$$K_{(1)} = (\overline{H_2S_2})/(\overline{HS})^2 = [\overline{H_2S_2}]\overline{y}_{H_2S_2}/[\overline{HS}]^2\overline{y}_{HS}^{\,2} = Q_{(1)}(\overline{y}_{H_2S_2}/\overline{y}_{HS}^{\,2}) \qquad \text{(III-2)}$$

In practically all of the studies, it is assumed that the activity coefficient quotient remains constant when $\overline{C}_{HS}$ is varied, and, indeed, constant values of the concentration quotient $Q_{(1)}$ are found over wide concentration ranges down to very low concentrations. Deviations from constancy have then been ascribed to further association (60D, 68K, 68L) rather than to the breakdown of the assumed constancy of the activity coefficient quotient.

Within the range where $Q_{(1)}$ is independent of $\overline{C}_{HS}$, it will be taken as a valid representation of the equilibrium constant $K_{(1)}$, since on extrapolation to $\overline{C}_{HS} = 0$, the activity coefficient quotient should extrapolate to unity, if the "wet" (water saturated) diluent is taken as the reference state. Reaction (III-1) has been studied by the distribution method, and where necessary, the acid dissociation and dimerization in the aqueous phase have been taken into account. It is surprising that for some diluents, very discordant values of $Q_{(1)}$ have been reported - for carbon tetrachloride the values range from 4×10^4

(63A) to 3×10^6 (60D) - but for others much better agreement among different authors has been achieved.

Some estimates can be discounted on the basis of obvious errors, e.g. neglect of important coexisting equilibria, but for others the only guide to selection of the most nearly correct value is the conformity of the value to a general trend with diluent properties. For the diluents which do not contain donor oxygen atoms, the solubility parameter is the property with which a correlation of log $Q_{(1)}$ values can most readily be justified (71K). Such a correlation is shown in Fig. III-1. Again, reasons for a parabolic course for this correlation have been offered (71K), and a curve which describes the data best obeys the following expression, as obtained from least squares fitting:

$$\log Q_{(1)} = 6.98\pm0.20 + (0.00\pm0.03)(\delta - 7.0) - (0.38\pm0.02)(\delta - 7.0)^2 \quad \text{(III-3)}$$

Accordingly, those data that do conform to that relationship, have been given greater weight, and the resulting selected values are shown in Table III-1. Still, some diluents fall completely outside the limits of the curve: chlorobenzene and bromoform. So does kerosene, if the material used was predominantly dodecane-like, and did not have sufficient aromatics to increase its solubility parameter from 7.8 (δ for dodecane) to 8.6 (δ = 9.2 for benzene), while chloroform is barely within the limits of the curve. Too much weight should, therefore, not be put on this (or any other) attempt to describe diverse diluents by a single parameter, and there is a finite probability that the correlation function, rather than the individual experimental log $Q_{(1)}$ values of the nonconforming solvents, is wrong.

As regards the solvents with oxygen donor atoms for which $Q_{(1)}$ values have been reported, the data are too few and too scattered to permit a clear correlation, at least with δ values. Where solvents have been studied by two authors, their values of log $Q_{(1)}$ are not too far apart. These solvents have been included in Fig. III-1 and Table III-1 for the purpose of comparison, while not very much can be said about the reliability of the constants.

Table III-1

Dimerization Equilibrium Quotients of (H)DBP in Organic Diluents at 25°C

Diluent	log $Q_{(1)}$	Refs.
Isooctane	7.11±0.05	70U
n-Hexane	6.95±0.05	60D,70U
n-Octane	6.95±0.05	65U,70U
Cyclohexane	6.50±0.05	68K
Carbon tetrachloride	6.21±0.05	68K
Toluene	5.46±0.05	68K
o-Xylene	5.29±0.05	68K
Benzene	4.84±0.05	59H,68K
Chloroform	4.50±0.10	59D,59H,60D,68K, 70U
Chlorobenzene	5.7 (?)	68K
Nitrobenzene	3.53±0.03	59H,68K
Bromoform	4.0 (?)	68K
Dibutyl ether	3.07±0.07	59D,68K
Di-iso-propyl ether	2.29±0.05	60D
Methyl isobutyl ketone	1.10±0.07	57D,68K
Tri-n-butyl phosphate	0.12 (?)	60D
Methyl isobutyl carbinol	<1.0 (?)	60D

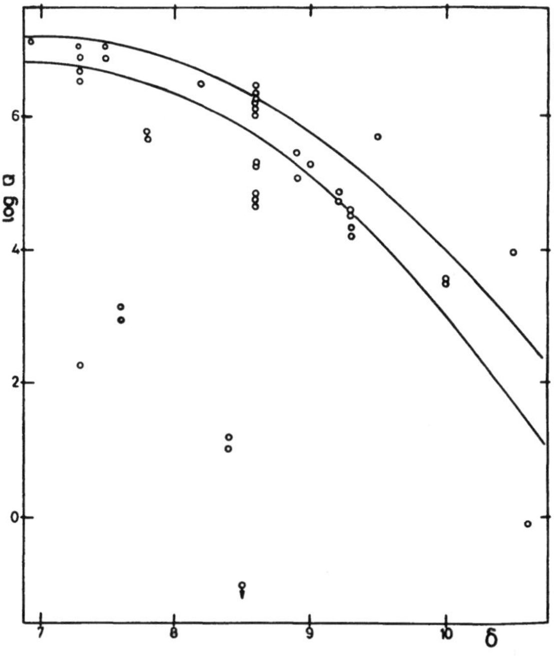

Figure III-1.

Dimerization equilibrium quotients of HS, log $Q_{(1)}$, as a function of the solubility parameter of the diluents, δ (71K). Continuous curves: $\log Q_{(1)} = 7.18 + 0.03\ (\delta - 7.0) - 0.36\ (\delta - 7.0)^2$ and $\log Q_{(1)} = 6.78 - 0.03\ (\delta - 7.0) - 0.40\ (\delta - 7.0)^2$. The diluents near the curves are from left to right: i-octane, hexane, n-octane, kerosene, c-hexane, CCl_4, toluene, o-xylene, benzene, chloroform, chlorobenzene, nitrobenzene and bromoform. Those at the bottom part are: di-i-propyl ether, dibutyl ether, methyl isobutyl ketone, methyl isobutyl carbinol, TBP.

References:

57D Dyrssen, D., Acta Chem. Scand., 11, 1771 (1957).

59D Dreze, Ph., Bull. Soc. Chim. Belges, 68, 674 (1959).

59H Hardy, C.J. and Scargill, D., J. Inorg. Nucl. Chem., 11, 128 (1959).

60D Dyrssen, D. and Liem, D.H., Acta Chem. Scand., 14, 1091 (1960).

63A Ampelogova, N.I., Radiokhimiya, 5, 562 (1963).

65U Ul'yanov, V.S. and Sviridova, R.A., Radiokhimiya, 7, 538 (1965).

68K Kolarik, Z., Hejna, J. and Pankova, H., J. Inorg. Nucl. Chem., 30, 2795 (1968).

68L Liem, D.H., Acta Chem. Scand., 22, 753 (1968).

70U Ul'yanov, V.S. and Sviridova, R.A., Radiokhimiya, 12, 47 (1970).

71K Kolarik, Z., Solv. Extr. Rev., 1, 1 (1971).

IV. Solvents: Di-n-butylphosphoric acid ((H)DBP), HS

Tri-n-butylphosphate (TBP), S'

Reactions:

$$1/2\ \overline{H_2S_2} + \overline{S'} = \overline{HS\cdot S'} \quad \text{(IV-1)}$$

$$\overline{H_2S_2} + \overline{S'} = \overline{(H_2S_2)S'} \quad \text{(IV-2)}$$

The interaction of HS with S' has been studied mainly in connection with its importance to the synergistic extraction of metal ions with these two solvents. Indeed, some of the data from which equilibrium constants for reactions (IV-1) and (IV-2) have been calculated are metal distribution data that only indirectly involve these constants, resulting in lowered accuracy. In studying the formation of the species $\overline{HS\cdot S'}$, many authors have considered instead of reaction (IV-1), the reaction

$$\overline{HS} + \overline{S'} = \overline{HS\cdot S'} \quad \text{(IV-3)}$$

Since in those diluents, for which comparative data exist, HS occurs predominantly as the dimer $\overline{H_2S_2}$, formulation as reaction (IV-1) is preferable to reaction (IV-3). Data for the latter have been standardized to the former, using the dimerization constants $K_{(4)}$ (or rather equilibrium quotients $Q_{(4)}$) for

$$\overline{HS} = 1/2\ \overline{H_2S_2} \quad \text{(IV-4)}$$

evaluated in Section III of this report.

The equilibrium constant for reaction (IV-1) is

$$K_{(1)} = (\overline{HS\cdot S'})/(\overline{S'})(\overline{H_2S_2})^{1/2} = [\overline{HS\cdot S'}]\overline{y}_{HS.S'}/[\overline{S'}][\overline{H_2S_2}]^{1/2}\overline{y}_{S'}\overline{y}_{H_2S_2}{}^{1/2} =$$

$$= Q_{(1)}(\overline{y}_{HS.S'}/\overline{y}_{S'}\overline{y}_{H_2S_2}{}^{1/2}) \quad \text{(IV-5)}$$

and for reaction (IV-2) it is

$$K_{(2)} = ((\overline{H_2S_2)S'})/(\overline{S'})(\overline{H_2S_2}) = [(\overline{H_2S_2)S'}]\overline{y}_{(H_2S_2S')}/[\overline{S'}][\overline{H_2S_2}]\overline{y}_{S'}\overline{y}_{H_2S_2} =$$

$$= Q_{(2)}(\overline{y}_{(H_2S_2S')}/\overline{y}_{S'}\overline{y}_{H_2S_2}) \qquad \text{(IV-6)}$$

If infinitely dilute solutions in the wet (saturated with water, in equilibrium with the standard ionic medium of 0.1 M) diluents are taken as standard reference states, the activity coefficient quotients in (IV-5) and (IV-6) approach unity as $\overline{C}_{HS}$ and $\overline{C}_{S'}$ approach zero in the diluents, thus the limits of $Q_{(1)}$ and of $Q_{(2)}$ (i.e. $Q^{\infty}_{(1)}$ and $Q^{\infty}_{(2)}$) should be constants.

The values of the dimerization and distribution equilibrium quotients for HS in n-hexane and n-octane are nearly identical, so that the interaction constant $Q_{(1)}$ (or $Q_{(2)}$) should be the same in these diluents, as well as in n-heptane. Kerosene might, however, give somewhat different results, particularly if it contains an appreciable fraction of aromatics. Data obtained from the distribution of HS between the diluent, containing varying low concentrations of S', and an aqueous solution, with all necessary side reactions taken into account (57D, 59H, 60D, 66L, 67K, 67L, 68L) are considered the most accurate. While reaction (IV-2) has been ignored in ref. (67M), the assumptions involved in considering the activities of HS, referred to pure HS as the standard state, have not been justified sufficiently in ref. (67M) to make them reliable. The resulting values of log $Q^{\infty}_{(1)}$ are about 0.14 units too large (compared with those considered reliable) for the diluents studied in this work, heptane and benzene. Data obtained from the synergistic extraction of thorium (64D) or uranium (63B) cannot be considered as highly accurate, though they do serve as checks on the general validity of the constants. The equilibrium quotients considered reliable are shown in Table IV-1, the limits of "infinite dilution" for this purpose being 10^{-4} M for HS and 3×10^{-2} M for S'. The observed stronger dependence of $Q^{\infty}_{(2)}$ than that of $Q^{\infty}_{(1)}$ on the nature of the diluents is reasonable in view of the bonding relationships involved. In reaction (IV-1), one (one half of the two) hydrogen bond between HS molecules is exchanged for one hydrogen bond between HS and S'. While the dimer H_2S_2 has no free donor oxygen atoms, both S' and the adduct HS·S' have one each, so that the interaction with the diluent should not change greatly.

In reaction (IV-2), if a hydrogen bond is not broken as it is unlikely to be, the donor oxygen atom of S' becomes involved in interaction with the dimer so that it loses its former interaction with the diluent.

There remains the problem of the interaction of HS with undiluted, water saturated S'. It is questionable whether only a single S' molecule would react under these circumstances, but on the other hand, the direct hydrogen bond HS·S' should be stronger than the dipole interactions expected when more than one S' molecule reacts. Thus reaction (IV-1) may still proceed as written, while in reaction (IV-2), where mainly dipole interactions are involved, more than one S' molecule would react. The equilibrium quotient $\log Q^o_{(1)} = 3.71$ refers to water-saturated S' as the standard state, using $C_{S'} = 3.43$ M (at 20°) and log Q (dimerization) = 0.12. However, $\log Q^o_{(1)}$ for undiluted S' is algebraically much larger than $\log Q^\infty_{(1)}$ in the diluents

$$\Delta \log Q_{(1)} = \log Q^o_{(1)} - \log Q^\infty_{(1)} \simeq 4.2 \qquad \text{(IV-7)}$$

than for a comparable reaction, e.g. for the extraction of nitric acid

$$\Delta \log Q = \log Q^o - \log Q^\infty \approx 0.8 \ . \qquad \text{(IV-8)}$$

There is therefore some doubt as to whether, indeed, only one S' molecule participates, and possibly $\log Q^o_{(1)}$ needs considerable revision.

Table IV-1

Equilibrium Constants for (H)DBP + TBP = adducts at 25°C

Diluent	log $Q^{\infty}_{(1)}$	log $Q^{\infty}_{(2)}$	Refs.
Hexane or octane	-0.54±0.05	1.27±0.03	66L,67L,68L
Kerosene	-0.07±0.05	0.56±0.1	59H,63B,64H
Carbon tetrachloride	-0.46±0.04	0.20±0.05	63B,66L,67K,67L,68L
Chloroform	-0.65±0.05		57D,64D
Benzene	-0.49±0.05		68L

References:

57D Dyrssen, D., Acta Chem. Scand., 11, 1771 (1957).

59H Hardy, C.J. and Scargill, D., J. Inorg. Nucl. Chem., 11, 128 (1959).

60D Dyrssen, D. and Liem, D.H., Acta Chem. Scand., 14, 1091 (1960).

63B Baes, C.F. Jr., Nucl. Sci. Eng., 16, 405 (1963).

64D Dyrssen, D. and Liem, D.H., Acta Chem. Scand., 18, 224 (1964).

66L Liem, D.H. and Dyrssen, D., Acta Chem. Scand., 20, 272 (1966).

67K Kuča, L., Coll. Czech. Chem. Comm., 32, 720 (1967).

67L Liem, D.H., in "Solvent Extraction Chemistry," eds. D. Dyrssen, J.-O. Liljenzin and J. Rydberg, North-Holland, Amsterdam, 1967, p. 305.

67M Mikhailov, V.A., Grigoreva, E.F. and Kharchenko, S.K., Radiokhimiya, 9, 149 (1967).

68L Liem, D.H., Acta Chem. Scand., 22, 753 (1968).

V. Solvent: Tri-n-butylphosphate (TBP) , S

Reaction: $S = \overline{S}$ (V-1)

The distribution equilibrium of S between organic and aqueous phases has been studied fairly thoroughly. The thermodynamic equilibrium constant for reaction (V-1), $K_{(1)}$ with pure S as the standard state, is, of course, unity. It is, however, of interest to calculate the equilibrium quotient for undiluted S, as well as the limits of the distribution constant at infinite dilution of S in a diluent and in water (or the standard free energy of transfer between the two liquids).

Undiluted S, in equilibrium with water, forms a two-phase system. The composition of the organic phase is (62D, 64H, 66R) at 25°C:

$$\overline{C}_S = 3.43 \pm 0.02 \text{ M} \qquad \overline{C}_{H_2O} = 3.58 \pm 0.02 \text{ M} \qquad \text{(V-2a)}$$

$$\overline{X}_S = 0.489 \pm 0.003 \qquad \overline{X}_{H_2O} = 0.511 \pm 0.003 \qquad \text{(V-2b)}$$

From the Gibbs-Duhem equation, applied to vapor-phase equilibration (66R) which is preferable to liquid-liquid equilibration with electrolyte solutions at different a_{H_2O} (64H), the activity of S at $a_{H_2O} = 1$ has been obtained (activity coefficients refer to pure S as the standard state):

$$\overline{a}_{(C\text{-scale})S} = 1.880 \pm 0.005 \text{ M} \qquad \overline{y}^o_S = 0.549 \pm 0.004 \qquad \text{(V-3a)}$$

$$\overline{a}_{(X\text{-scale})S} = 0.514 \pm 0.002 \text{ M} \qquad \overline{f}^o_S = 1.054 \pm 0.007 \qquad \text{(V-3b)}$$

The equilibrium aqueous phase is so dilute in S that $a_{H_2O} = 1$, $X_{H_2O} = 1.000$ and $C_{H_2O} = 55.3$ are maintained. The solubility of S in water has been measured over a temperature range, and found to decrease with increasing temperatures. Since centrifugation has been used at lower temperature solubility measurements (16-22°C) in an uncontrolled manner (56A, 63B), the lower results may reflect heating during centrifugation (60H), since an increase of 8° brings these points neatly into place with those obtained at higher temperatures, and on a curve which includes very low (5°, 13°) and high (50°) temperature data also. At 25° (cf. also 59B, 62H) the best value for the aqueous solubility of S is

$$C_S = (1.561 \pm 0.017)10^{-3}\ M \qquad X_S = (2.80 \pm 0.03)10^{-5} \qquad \text{(V-4)}$$

From eq. (V-2) to (V-4) the distribution equilibrium quotient is obtained

$$\log Q_{(1)}\ \text{(C-scale)} = 3.341 \pm 0.005 \qquad \text{(V-5a)}$$

$$\log Q_{(1)}\ \text{(X-scale)} = 4.241 \pm 0.005 \qquad \text{(V-5b)}$$

The activity coefficients in the aqueous phase, referred to pure S as the standard state, are

$$y_S^o = (1.20 \pm 0.01)10^3 \qquad f_S^o = (1.83 \pm 0.02)10^4 \qquad \text{(V-6)}$$

for the molar and mole-fraction scales respectively.

Turning now to dilute solutions of S in diluents, and the infinitely dilute solution as the reference state for activity coefficients, it has been argued that the aqueous solubilities (eq. (V-4)) are so low, that $y_S = 1$ should apply to them. Indeed, in the range 8×10^{-4} to 1.25×10^{-2} M S in kerosene, the distribution coefficients of S are essentially constant, (53K, 56A) so that activity coefficients in both phases can be taken as equal unity. Increasing distribution coefficients at higher concentrations should then reflect increase in $\overline{y}_S$. The results, expressed as

$$Q_{(1)}^{\infty} = \lim\ (\overline{C}_S \rightarrow 0) Q_{(1)} \qquad \text{(V-7)}$$

are shown in Table V-1. They were obtained mainly for <0.03 M solutions

(53K,56A) but for several solvents, high concentrations 0.18-0.21 M (53J, 61M, 62S) had to be used, corrected for $\overline{y}_S$ by comparison with solvents for which data at low concentrations are available.

The activity coefficients vary in a parallel manner for the solvents concerned (56A). The temperature variation of $Q^{\infty}_{(1)}$ is ignored for the range 20-25°.

Table V-1

Distribution constants for TBP = $\overline{\text{TBP}}$ at 20-25°

Solvent	log $Q^{\infty}_{(1)}$	References
Kerosene	2.40±0.03	53J,53K,56A,62S
Dodecane	2.46±0.04	53J,61M
Heptane	2.66±0.03	56A,62S
Hexane	2.80±0.03	56A,62S
Cyclohexane	2.88±0.03	53J,56A,62S
Benzene	3.80±0.03	53J,56A
Carbon tetrachloride	3.58±0.02	53J,62S
Toluene	3.9 [x)]	62S
Chloroform	4.9 [x)]	62S

[x)] These values are rather uncertain since they disagree with the values given by the other authors (53J,56A), but were selected since they fit better to the presumed solvent effects.

References:

53J Johnson, W.F., Dillon, R.L., U.S. Report HW 29086 (1953).

53K Kennedy, J., Grimely, S.S., British Report AERE CE/R 1283 (1953).

56A Andelin, R.L., Anderson, E.L. and McVey, H.W., USAEC Report IDO-14381 (1956).

59B Baldwin, W.H., Higgins, C.E. and Soldano, B.A., J. Phys. Chem., 63, 118 (1959).

60H Higgins, C.E. and Baldwin, W.H., Anal. Chem., 32, 326 (1960).

61M Marcus, Y., J. Phys. Chem., 65, 1647 (1961).

62D Davis, W., Nucl. Sci. Eng., 14, 159 (1962).

62H Higgins, C.E. and Baldwin, W.H., J. Inorg. Nucl. Chem., 24, 415 (1962).

62S Siekierski, S., J. Inorg. Nucl. Chem., 24, 205 (1962).

63B Burger, L.L., Nucl. Sci. Eng., 16, 428 (1963).

64H Hardy, C.J., Fairhurzt, D., McKay, H.A.C. and Wilson, A.M., Trans. Faraday Soc., 60, 1626 (1964).

66R Roddy, J.W. and Mrocheck, J., J. Inorg. Nucl. Chem., 28, 3019 (1966).

VI. Solvent: Tri-n-butylphosphate (TBP), S

Ligand: Water (H_2O), W

Reaction: $W + \overline{S} = \overline{W \cdot S}$ (VI-1)

This reaction is the major one at low water and S concentrations, but other products besides $\overline{W \cdot S}$ are present at higher concentrations. At high water and low S concentrations there is good evidence for "dimeric" water bound to S, i.e., $\overline{W_2 \cdot S}$ (67J, 68R, 69B), while at low water-high S concentrations species such as $\overline{W \cdot S_2}$ are formed (64H,68N). Still more complex species were reported for solutions containing high concentrations of both S and water, such as $\overline{W_2S_2}$ or even the corresponding decamer (68R). Undiluted, water-saturated S has the composition $\overline{W_x \cdot S}$ where x is near unity, but significantly different from it: $x = 1.045 \pm 0.001$ at 25° (62D, 64H) and varies with temperature, from 1.13 at 0° to 0.91 at 60° (64H), so that several species must exist there in equilibrium, besides the $\overline{W \cdot S}$.

Most of the information on reaction (VI-1) has been obtained from equilibrations of dilute solutions of S in inert solvents with water or dilute aqueous solutions. The equilibrium constant expression therefore uses a mixed concentration scale: molar for $\overline{S}$ and $\overline{W \cdot S}$ and mole fraction for W (i.e. $a_W = 1$ in pure water).

$$K_{(1)} = (\overline{W \cdot S})/(\overline{S})(W) = \lim (a_W \rightarrow 1)(\overline{WS})/(\overline{S}) \qquad \text{(VI-2)}$$

Usually the concentration quotient

$$Q_{(1)} = [\overline{W \cdot S}]/[\overline{S}]a_W = K_{(1)}(\overline{y}_S/\overline{y}_{WS}) \qquad \text{(VI-3)}$$

is calculated, with the assumption that $(\overline{y}_S/\overline{y}_{WS})$ would be a constant for dilute solutions of S in the diluent. Straight line plots of $[\overline{W \cdot S}]$ vs. $[\overline{S}]$ confirm the constancy of the activity coefficient qoutient in dilute solutions ($\overline{C}_S$ < 0.07 M for iso-octane, <0.10 M for CCl_4, <0.20 M for 1,2 dichloroethane (63W, 69B)).

Values of $\lim Q_{(1)}(\overline{C}_S \rightarrow 0)$ are shown in Table VI-1. These are valid for saturated solutions of water in the inert diluents, i.e. equilibrium with water at $a_W = 1$, as the standard state.

In order to refer to pure dry inert diluent as the standard state, infinite dilution of both S and W is required. NMR chemical shift data in CCl_4 solutions dilute in water, $\overline{X}_W < 2\times10^{-3}$ were interpreted according to the reaction

$$\overline{W} + \overline{S} = \overline{W \cdot S} \qquad \text{(VI-4)}$$

with

$$K^{\infty}_{(4)} = \lim (\overline{C}_W \rightarrow 0,\ \overline{C}_S \rightarrow 0)[\overline{W \cdot S}]/[\overline{S}][\overline{W}] \qquad \text{(VI-5)}$$

In order to convert this constant to that corresponding to reaction (VI-1), $K^{\infty}_{(4)}$ must be multiplied with the Nernst distribution constant for water, i.e. the equilibrium constant for

$$W = \overline{W} \qquad \text{(VI-6)}$$

$\lim\,(a_W \to 0)\,[\overline{W}]/a_W$ being assumed to be the same as $[\overline{W}]_{CCl_4}$ at $a_W = 1$, since the solutions, which are so dilute, can be assumed to be ideal. The value

$$[\overline{W}] = (7.15 \pm 0.53)\times 10^{-3}\ \text{M} \qquad \text{(VI-7)}$$

obtained from extrapolation of distribution data with $\overline{C}_S \to 0$ (68R) seems to be the most reliable. The original NMR data (68N) were expressed in terms of mole fractions, and can be converted to the molar scale by the use of CCl_4 densities at the temperatures of measurement: 10, 21, 35 and 45°,

$$d_{CCl_4} = 1.63255 - 1.9110\times 10^{-3}t - 0.690\times 10^{-6}t^2 \qquad \text{(VI-8)}$$

(Intl. Crit. Tables, 3, 33) and interpolation to 25°. Thus the value of

$$K^{\infty}_{(1)} = \lim\,(\overline{C}_W \to 0,\ \overline{C}_S \to 0)\ [\overline{W\cdot S}]/[\overline{S}]a_W =$$

$$= K^{\infty}_{(1)}\ (\text{x-scale})\,(M_{CCl_4}/1000d_{CCl_4})\,[\overline{W}]_{CCl_4} \qquad \text{(VI-9)}$$

where $M_{CCl_4} = 153.82$ is the gram-molecular weight, is given in Table VI-1.

The corresponding value of $K^{\infty}_{(1)}$ for kerosene diluent is more difficult to calculate. One can start from the value of $\lim\,(a_W \to 0)\ Q_{(1)} = 0.31 \pm 0.02$ at the given, finite S concentration of $\overline{C}_S = 0.732$ M (20% by volume in kerosene of d = 0.778), obtainable from the slope of a $\overline{C}_W$ vs. a_W plot (64H). The activity coefficient $\overline{y}_S$ is now evaluated, referred to $\overline{C}_S \to 0$, from vapor pressure data for dry S in a hydrocarbon diluent. For lack of data for kerosene solutions, data for hexane solutions will be used (63P):

$$\log \overline{f}_S = (a + \tfrac{3}{2}b + 2c)\overline{X}_{HxH}^{\,2} - (b + \tfrac{8}{3}c)X_{HxH}^{\,3} + c\overline{X}_{HxH}^{\,4}\ , \qquad \text{(VI-10)}$$

with a = 1.39, b = -2.15 and c = 1.00. The 0.732 M solution contains 622.4 g diluent per liter, so that its mole fraction, if it were hexane, is 0.908. The activity coefficient of trace S, at $\bar{x}_{HxH} = 1$, is $\log \bar{f}^o_S = 0.648$, and that of 0.732 M S, referred to the trace as standard state is

$$\log \bar{f}'_S = \log (\bar{f}_S/\bar{f}^o_S) = 0.429 - 0.648 = -0.219 \qquad \text{(VI-11)}$$

on the mole fraction scale. Converted to the molar scale, $\bar{y}_S = 0.618$, and this with $\lim Q_{(1)}$ gives the value of $K^{\infty}_{(1)}$ shown in Table VI-1. The value of $\bar{y}_S$ in dry kerosene, calculated here, is considerably higher than the value in wet kerosene, 0.325, interpolated from a graph (56A) for the same concentration, as is expected. The value of $K^{\infty}_{(1)}$ in kerosene is much larger than the value in CCl_4 (Table VI-1), probably because the polarizability of the latter solvent does not lead to such a tight mutual binding of the polar water and the TBP as does the more inert kerosene.

The equilibrium constant for reaction (VI-1) can be calculated also for undiluted S from data at variable a_W, provided undiluted S is considered as monomeric, and only one S is involved in binding the W. In fact, a linear relation $\bar{x}_W = K^o_{(1)} a_W$ is obtained at $a_W < 0.5$ from both vapor pressure (67J) and distribution (64H) data, when results are corrected for $\overline{W_2 \cdot S}$ formation. On the mole fraction scale, the former data lead to

$$K^o_{(1)} = \lim (x_S \rightarrow 1,\ a_W \rightarrow 0)\bar{x}_W/a_W \qquad \text{(VI-12)}$$

and the value of $K^o_{(1)}$ shown in Table VI-1. The latter data were presented on the molar scale, and conversion of $\bar{C}_W$ to $\bar{x}_W$ is effected by dividing by $\bar{C}_S = 3.655$ in undiluted S (62D). However, the value found is high by a factor of 1.30. The possibility that $\bar{C}_W$ has been overestimated in this work, leading to a high $K^o_{(1)}$, must however be considered, since the Gibbs-Duhem estimates of $\bar{a}_S$ are correspondingly low, and at the limit of $a_W = 1$, where data for comparison exist, the ratio $\overline{a}_S$(distribution)/$\overline{a}_S$(vapor pressure) = 0.63. There is also considerable scatter in the data at low $\overline{C}_W$, so that on the whole, the vapor pressure data (67J) seem to be more reliable. Since, however, the undiluted S probably does not contain $\overline{W \cdot S}$ as the main species even if $a_W \to 0$, but rather a mixture of $\overline{WS_i}$ species, the value of $K^o_{(1)}$ calculated above and shown in Table VI-1 does not strictly pertain to reaction (VI-1).

Table VI-1

Equilibrium Constants for $H_2O + \overline{TBP} = \overline{H_2O \cdot TBP}$ at 25°C

Solvent	log lim $Q_{(1)}$ ($\overline{C}_S \to 0$)	log $K^{\infty}_{(1)}$	log $K^o_{(1)}$ (x-scale)
Kerosene	-0.92±0.04	-0.30±0.03	
(Refs.)	(56A,61B,62D,64H,69B)	(64H,63P)	
Benzene	-0.41±0.04		
(Refs.)	(61C)		
CCl_4	-0.82±0.02	-1.36±0.03	
(Refs.)	(63W,67K,68R)	(68N)	
CH_2ClCH_2Cl	-0.36±0.04		
(Refs.)	(69B)		
Undiluted			-0.41±0.01
(Refs.)			(67J)

References:

56A Alcock, K., Grimely, S.S., Healy, T.V., Kennedy, J. and McKay, H.A.C., Trans. Faraday Soc., 52, 39 (1956).

61B Brauer, E. and Högfeldt, E., J. Inorg. Nucl. Chem., 23, 115 (1961).

61C Chaikhorskii, A.A., Vdovenko, V.M., Efimova, K.I. and Belov, L.M., Radiokhimiya, 3, 295 (1961).

62D Davis, W., Nucl. Sci. Eng., 14, 159 (1962).

63P Pushlenkov, M.F. and Shuvalov, O.N., Radiokhimiya, 5, 543 (1963).

63W Whitney, D.C. and Diamond, R.M., J. Phys. Chem., 67, 209 (1963).

64H Hardy, C.J., Fairhurst, D., McKay, H.A.C. and Wilson, A.M., Trans. Faraday Soc., 60, 1626 (1964).

67J Johnson, J.R., Christian, S.D. and Affsprung, H.E., J. Chem. Soc., 1967, 1924 (1967).

67K Kuča, L., Coll. Czech. Chem. Comm., 32, 720 (1967).

68N Nishimura, S., Ke, C.H. and Li, N.C., J. Am. Chem. Soc., 90, 234 (1968).

68R Roland, G. and Duyckaerts, G., Spectrochim. Acta, 24A, 529 (1968).

69B Bucher, J.J. and Diamond, R.M., J. Phys. Chem., 73, 675 (1969).

VII. Solvent: Tri-n-butylphosphate (TBP) , S

Ligand: Nitrate (NO_3^-) , A^-

Reaction: $H^+ + A^- + \overline{S} = \overline{HAS}$ (VII-1)

This reaction is the major one in this system with both undiluted S as well as with its solutions in inert diluents, provided the HA concentration in the equilibrium aqueous phase is below 7 M. (56A, 56F, 57A, 58P, 58S, 60A, 62D, 63N, 64Z, 65K, 65S, 69S)

Above this concentration, more HA enters the organic phase than corresponds to $\overline{HAS}$. This has been interpreted as physical dissolution of undissociated HA (56A)

$$HA = \overline{HA} \qquad \text{(VII-2)}$$

or formation of higher complexes, in particular (56Aa, 57A, 58F, 60A, 60S, 64S)

$$H^+ + A^- + \overline{HAS} = \overline{(HA)_2S} \qquad \text{(VII-3)}$$

Most authors agree that $\overline{HAS}$ is essentially unhydrated, but some discuss species such as $\overline{HAS \cdot H_2O}$ (59D) or $\overline{H(H_2O)_4AS}$ (64E). However, the latter has been found to be energetically unfavored (67B). The effect of water on the dissociation of extracted HA

$$\overline{HA} = \overline{H^+} + \overline{A^-} \qquad \text{(VII-4)}$$

without identified species, has however been noted (67B), and the heat effects of the dissociation reaction

$$\overline{HAS} = HA + \overline{S} \qquad \text{(VII-5)}$$

where HA is either the pure liquid or gaseous nitric acid, has been determined (67B).

Returning now to the major reaction (VII-1), this has received also the alternative formulation (49M, 55G, 56A, 56H, 57K, 58C, 59S, 60S, 67B)

$$HA + \overline{S} = \overline{HAS} \qquad \text{(VII-6)}$$

Reactions (VII-1) and (VII-6) obviously differ only by the association of nitric acid in the aqueous phase

$$H^+ + A^- = HA \qquad \text{(VII-7)}$$

In the following analysis, reaction (VII-1) will be considered, and data reported according to reaction (VII-6) will be converted to the form of reaction (VII-1) by the use of $K_{(7)} = 0.053 \pm 0.007$, the most recent critically evaluated constant (64D).

A critical evaluation of distribution data published prior to 1961 for undiluted S and for solutions of S in saturated hydrocarbon diluents (i.e. hexane and odorless kerosene) has been published (62Da). This work and a consideration of later publications lead to the conclusion that in reaction (VII-1), the symbol $\overline{S}$ stands for the average hydrated species of TBP, and the species $\overline{HAS}$ for the average hydrated species of $HNO_3 \cdot TBP$, where the degree of hydration of the former is appreciably higher than that of the latter. However, the degree of hydration is a function of $\overline{C}_S$ and of $\overline{C}_{HA}$, and can be expressed best by the activity coefficients $\overline{y}_S$ and $\overline{y}_{HAS}$. The non-formation of one definite S-water species, which is now generally accepted, makes the alternative interpretation, in terms of species containing definite stoichiometries of water, and of water set free during the reaction, much less attractive.

The equilibrium constant for reaction (VII-1) is

$$K_{(1)} = (\overline{HAS})/(\overline{S})(H^+)(A^-) = [\overline{HAS}]\overline{y}_{HAS}/[\overline{S}]\overline{y}_S[H^+][A^-]y^2_{\pm HA} \qquad \text{(VII-8)}$$

Many authors report the concentration quotient

$$Q_{(1)} = [\overline{HAS}]/[\overline{S}][H^+][A^-] = K_{(1)}(\overline{y}_S/\overline{y}_{HAS})y^2_{\pm HA} \qquad \text{(VII-9)}$$

Surprisingly, $Q_{(1)}$ is a constant, within limits, independent of $\overline{C}_{HA}$ in the range 0.2 to 3.0 M, of $\overline{C}_S$ below 1 M, and of the temperature in the range 20-70°C (60A). It is dependent on the solvent, however, as shown in Table VII-1. In the evaluation of $Q_{(1)}$, the change in [S], due to volume changes caused by the extraction reaction, has been taken into account. The apparent constancy of $Q_{(1)}$ is due to compensation effects on the activity coefficient term on the right hand side of (VII-9). Attempts to correct $Q_{(1)}$ for aqueous phase activity coefficients lead to values of

$$Q'_{(1)} = Q_{(1)}/y^2_{\pm HA} = [\overline{HAS}]/[\overline{S}][H^+][A^-]y^2_{\pm HA} = K_{(1)}(\overline{y}_S/\overline{y}_{HAS}) \qquad \text{(VII-10)}$$

which is found to vary with $\overline{C}_{HA}$ and $\overline{C}_S$. However, in the region where $y_{\pm HA}$ varies only slowly with $\overline{C}_{HA}$ (i.e. 0.3 to 1.5 M HA, where $y_{\pm HA} = 0.75\pm0.02$), the approximate constancy of $Q_{(1)}$ has as a corrolary the approximate constancy of $Q'_{(1)}$ as shown in Table VII-1.

In order to evaluate the activity coefficients, the standard states must be defined. If infinite dilution of all solutes in the organic phase is used (neglecting dissolved water, in equilibrium with the aqueous phase), the resulting constant is

$$K^{\infty}_{(1)} = \lim\ (\overline{C}_S \rightarrow 0,\ \overline{C}_{HA} \rightarrow 0)Q_{(1)} \qquad \text{(VII-11)}$$

If only C_{HA} is extrapolated to zero, at finite $\overline{C}_S$, an equilibrium quotient is obtained

$$Q''_{(1)} = \lim\ (C_{HA} \rightarrow 0)Q_{(1)} = K_{(1)}\overline{y}_S/\overline{y}_{HAS\ (trace)} \qquad \text{(VII-12)}$$

which can be a constant (independent of $\overline{C}_S$) only if $\overline{y}_{HAS\ (trace)}$ varies with $\overline{C}_S$ in the same manner that $\overline{y}_S$ does. This has not been tested extensively, because of the lack of data at very low C_{HA}, but in the cases of kerosene diluent (59J) and undiluted S (64Z) seems to lead to reasonable values $Q''_{(1)} \approx K_{(1)}$, Table VII-1.

The double extrapolation (VII-11) has been done for only a few solvents: kerosene (Amsco 125-82) and hexane (62Da) and CCl_4 (63N), Table VII-1. Alternatively, attempts to evaluate $\overline{y}_{HAS}$ and $\overline{y}_S$ at finite concentrations, as functions of $\overline{C}_S$ and $\overline{C}_{HA}$, have also been made. The functions have been written in terms of "salting coefficients" (56A, 61M, 62D, 62Da, 63N) or some other dependence on the concentration (61O). The evaluation of $\overline{y}_{HAS}$ is in any case based on an extrapolation of distribution data, but $\overline{y}_S$ can, in principle, be obtained independently, from partial vapor pressure data. These coefficients permit the evaluation of $K_{(1)}$ at any concentration of S, and yield, in fact, a constant equivalent to $Q''_{(1)}$.

Another definition of the standard state is in terms of undiluted but hydrated S, in equilibrium with pure water: the constant $K^o_{(1)}$ is defined as $K_{(1)}$ of eq. (VII-8) where $\lim \overline{y}_S([S] \to [S]^\circ) = 1$. Here

$$[S]^\circ = 1000\rho/(266.32 + 18.01x) \qquad \text{(VII-13)}$$

and x is the molar ratio of water to S in the undiluted hydrated S at the equilibrium temperature, and ρ is the density of this phase as C_{HA} goes to zero. The problem here is that although $\lim \overline{y}_{\pm HA}(C_{HA} \to 0) = 1$, the same does not necessarily hold for $\overline{y}_{HAS}$, which assumes a definite, but unknown value, related to the standard free energy of transfer of HA from its infinitely dilute solution in water to a corresponding solution in undiluted hydrated S. Thus only

$$Q^o_{(1)} = K^o_{(1)}\overline{y}^o_{HAS} = \lim\, [C_{HA} \to 0,\ [S] \to [S]^\circ]\, Q_{(1)} \qquad \text{(VII-14)}$$

can be determined. Extrapolations of $Q_{(1)}$ for undiluted S measured over a range of C_{HA} (58Sa, 59D, 61O, 62D, 62Da, 64Z) to $C_{HA} = 0$ yields the value given in Table VII-1 for log $Q''_{(1)}$ and undiluted S. Extrapolation of the values of log $Q''_{(1)}$ at given S concentrations to $[S]^\circ$ yields the value of log $Q^o_{(1)}$ given in Table VII-1 (62Da). There should be agreement between $Q''_{(1)}$ (undiluted S) and $Q^o_{(1)}$, and the value of $Q^o_{(1)}$ given in Table VII-1 should be considered the more reliable estimate for this quantity, since the extrapolated $Q''_{(1)}$ at low C_{HA} is highly curved, while that for $Q^{o}_{(1)}$ is linear, although the data used for the latter show high scatter.

Table VII-1

Equilibrium Constants for $H^{+} + NO_3^{-} + \overline{TBP} = \overline{HNO_3 \cdot TBP}$ at 20-25°C

Solvent	Kerosene	Benzene	CCl_4	Undiluted
$\log Q_{(1)}$	-0.67±0.06	-0.91±0.08	-1.09±0.13	
(Refs.)	(56C,57S,58P, 58S,60A,63S)	(56F,58S)	(58S,64R,64S)	
$\log Q'_{(1)}$	-0.42±0.06	-0.66±0.08	-0.84±0.13	
(Refs.)	(57A,58S)	(65S,69F)	(58S,64R,64S)	
$\log Q''_{(1)}$	-0.4±0.2			0.03±0.03
(Refs.)	(59J)			(58S,59D,610, 62D,64Z)
$\log K^{\infty}_{(1)}$	-0.64±0.01		-0.60±0.05	
(Refs.)	(62Da)		(63N)	
$\log Q^{o}_{(1)}$				0.17±0.02
(Refs.)				(610,62Da)

References:

49M Moore, R.L., USAEC Report AECD - 3196 (1949).

55G Geary, T.V., British Report AEAR - 8142 (1955).

56A Alcock, K., Grimely, S.S., Healy, T.V., Kennedy, J. and McKay, H.A.C., Trans. Faraday Soc., 52, 39 (1956).

56Aa Andelin, T.L., Anderson, E.L. and McVey, H.W., USAEC Report IDO-14381 (1956).

56F Fomin, V.V. and Maiorova, E.P., Zh. Neorg. Khim., 1, 1703 (1956).

56H Healy, T.V. and Brown, P.E., British Report AERE C/R 1970 (1956).

57A Allen, R.J. and Desesa, M.A., Nucleonics, 15(9) 88 (1957).

57K Karpacheva, S.M., Khorkhorina, L.P. and Rozen, A.M., Zh. Neorg. Khim., 2, 1441 (1957).

57R Rozen, A.M., Zh. Neorg. Khim., 2, 1956 (1957).

58C Codding, J.W., Jr., USAEC Report IDO-14454 (1958).

58F Fomin, V.V. and Maiorova, E.P., Zh. Neorg. Khim., 3, 540 (1958).

58P Povitskii, N.S., Solovkin, A.S. and Shilin, I.V., Zh. Neorg. Khim., 3, 222 (1958).

58S Shevchenko, V.B., Povitskii, N.S., Solovkin, A.S., Shilin, I.V., Lunichkina, K.P. and Tsvetkova, Z.N., Zh. Neorg. Khim., 3, 2109 (1958).

59D Damiani, L. and Fattore, V., Energia Nucleare (Milan), 6, 793 (1959).

59J Jury, S.H. and Whatley, R.F., USAEC Report CF-59-8-122 (1959).

59S Sheka, Z.A. and Kriss, E.E., Zh. Neorg. Khim., 4, 2505 (1959).

60A Adamskii, N.M., Karpacheva, S.M., Melnikov, I.N. and Rozen, A.M., Radiokhimiya, 2, 13 (1960).

60S Solovkin, A.S., Thesis, Moscow, 1960 (quoted in 64S).

61M Marcus, Y., J. Phys. Chem., 65, 1641 (1961).

61O Olander, D.R., Donadieu, L. and Benedict, M., A.I.C.E.J., 7, 152 (1961).

62D Davis, W., Nucl.Sci. Eng., 14, 174 (1962).

62Da Davis, W., Nucl. Sci. Eng., 14, 159 (1962).

63N Nishi, T. and Asano, M., Kogyo Kagaku Zasshi, 66, 1424 (1963).

64D Dyrssen, D. and Liem, D.H., Acta Chem. Scand., 18, 224 (1964).

64E El-Guebeily, M.A., Hallaba, E. and Azzam, R.A.I., Proc. Int. Conf. Peaceful Uses At. Energy, Geneva, 10, 452 (1964).

64S Solovkin, A.S., Zh. Neorg. Khim., 9, 746 (1964).

64Z Zemlyanukhin, V.I., Savoskina, G.P. and Pushlenkov, M.F., Radiokhimiya, 6, 714 (1964).

65K Korovin, S.S., Lebedeva, G.N., Dedich, K., Reznik, A.M. and Rozen, A.M., Zh. Neorg. Khim., 10, 518 (1965).

65S Shvedov, V.P. and Orlov, Yu.F., Zh. Neorg. Khim., 10, 2774 (1965).

67B Biddle, P., Coe, A., McKay, H.A.C., Miles, J.H. and Waterman, M.J., J. Inorg. Nucl. Chem., 29, 2615 (1967).

69S Shehata, M.K.K., and Aten, A.H.W., Jr., J. Inorg. Nucl. Chem., 31, 239 (1969).

VIII. Solvent: Tri-n-butylphosphate (TBP), S

Ligand: Chloride (Cl^-), A^-

Reactions: $H^+ + A^- + \overline{S} = \overline{HAS}$ (VIII-1)

$$H^+ + A^- + nH_2O + m\overline{S} = \overline{H(H_2O)_nAS_m} \quad \text{(VIII-2)}$$

The hydrochloric acid - TBP system is not very stable chemically, and eventually extensive hydrolysis occurs. Also, the extractability of hydrochloric acid is rather low, so that either high $\overline{S}$ concentrations or high HA concentrations are required in order to have conveniently measureable concentrations of HA in the organic phase. This explains the scarcity of reliable data for the calculation of the equilibrium constant of reaction (VIII-1).

Reaction (VIII-1) does express the major reaction when both HA and S are dilute in their respective phases. The equilibrium constant is

$$K_{(1)} = [\overline{HAS}]\overline{y}_{HAS}/[H^+][A^-]y_{\pm HA}^2[\overline{S}]\overline{y}_S = \lim(\overline{C}_S \rightarrow 0,\ C_{HA} \rightarrow 0)Q_{(1)}(\overline{y}_{HAS}/\overline{y}_S) \qquad \text{(VIII-3)}$$

where

$$Q_{(1)} = \overline{C}_{HA}/(\overline{C}_S - \overline{C}_{HA})a_{\pm HA}^2 \qquad \text{(VIII-4)}$$

Since at the limit of infinite dilution $\overline{y}_{HAS}/\overline{y}_S = 1$,

$$Q^{\infty}_{(1)} = \lim Q_{(1)}(\overline{C}_S \rightarrow 0,\ C_{HA} \rightarrow 0) = K_{(1)}. \qquad \text{(VIII-5)}$$

It may be expected that the ratio of activity coefficients $\overline{y}_{HAS}/\overline{y}_S$ approaches unity faster than does $y_{\pm HA}$, so that with relatively dilute solutions of S, rather concentrated aqueous solutions of HA can be handled and still not be too far away from the limit of infinite dilution in the organic phase.

Data for dilute benzene (65L), xylene (63A), kerosene (60A) and carbon tetrachloride (68M) solutions of S have been reported, from which equilibrium constants have been calculated. The former, unfortunately cannot be correct, since for a fourfold increase in aqueous HA concentration from 1 M to 4 M, corresponding to about a hundredfold increase in $a_{\pm HA}$, there is only a maximal twofold increase in the distribution coefficient of HA. The data for kerosene are unfortunately too sketchy to permit evaluation. The other two solvents, however, yield reasonable results, and the log $Q^{\infty}_{(1)}$ values are shown for xylene and carbon tetrachloride in Table VIII-1.

Although some water may participate in the extraction reaction, e.g. reaction (VIII-2) with m = 1, at the limit of infinite dilution $a_{H_2O} = 1$, and the mixed concentration scale (X-scale for water, C-scale for the other reactants) constant $K_{(2)}$ should equal $K_{(1)}$. It should be noted that the constants refer to water-saturated solvents (xylene and carbon tetrachloride) as the standard states.

For undiluted S, however, the role of water may be more important. There is good evidence that three moles of water are coextracted with the HA (59B, 59I, 66K) , but even here the stoichiometry has been variously reported between two and four (59I, 63A, 67B). As regards the number of moles of S reacting, this again has been controversial, and numbers from one to three have been reported (66L, 67B). If, however, undiluted, water-saturated S is considered the standard state, the activity of S in this state is a constant (a_S = 1.880 ±0.005 on the C-scale and 0.514±0.002 on the X-scale) and the equilibrium constant for reaction (VIII-2) must be divided by $\bar{a}_S^{\,m}$ if m is known. For this standard state, a_{H_2O} = 1 (X-scale) in any case, so the value of n is immaterial.

In the following, the equilibrium quotient

$$Q_{(2)} = \bar{C}_{HA}/a_{\pm HA}^{\;2} \qquad \text{(VIII-6)}$$

will be considered, and its limiting value as

$$Q^{\infty}_{(2)} = \lim\,(C_{HA} \rightarrow 0,\ \bar{C}_S \rightarrow 1000\rho/(266.32+18.01\ xH_2O))Q_{(2)} \qquad \text{(VIII-7)}$$

where x is the molar ratio of water to S in undiluted, hydrated S and ρ is the density of this phase, will be taken as the equilibrium constant, setting $\bar{a}_S$ = 1 arbitrarily. The data of several authors (59B, 59I, 60H) can be utilized, since they pertain to sufficiently low C_{HA} (or $\bar{C}_{HA}$) to permit the extrapolation for $Q^{\infty}_{(2)}$. Some other data (66K, 66L, 68S) cannot, however, be used, since they were obtained at too high values of C_{HA}, and yield small values $Q_{(2)}$, which cannot be extrapolated back. The activity coefficient $\bar{y}_{HA}$ has been reported as a function of $\bar{C}_{HA}$, but although two reported functions disagree (59B, 60H), they extrapolate to unity at the limit of $Q^{\infty}_{(2)}$ so that their disagreement is immaterial. The data have been corrected to the molar scale where necessary (59B). The values obtained are shown in Table VIII-1.

Table VIII-1

Equilibrium Constants for $H^+ + Cl^- + \overline{TBP(hydrated)} = \overline{HCl \cdot TBP(hydrated)}$

Solvent	Constant	Refs.
Xylene	$\log Q^{\infty}_{(1)} = -4.71 \pm 0.03$	63A
Carbon tetrachloride	$\log Q^{\infty}_{(1)} = -4.65 \pm 0.03$	68M
Undiluted	$\log Q^{\infty}_{(2)} = -1.68 \pm 0.03$	59B,59I[a],60H

[a]Data for 21° but the precision obtained makes them indistinguishable from the data at 25°C.

References:

59B Baldwin, W.H., Higgins, C.E. and Soldano, B.A., J. Phys. Chem., 63, 118 (1959).

59I Irving, H. and Edgington, D.N., J. Inorg. Nucl. Chem., 10, 306 (1959).

60A Adamskii, N.M., Karpacheva, S.M., Melnikov, I.N. and Rozen, A.M., Radiokhimiya, 2, 13 (1960).

60H Hesford, E. and McKay, H.A.C., J. Inorg. Nucl. Chem., 13, 156 (1960).

63A Apraksin, L.A., Korovin, S.S., Reznik, A.M. and Rozen, A.M., Zh. Neorg. Khim., 8, 237 (1963).

65L Levin, V.I. and Kozlova, M.D., Radiokhimiya, 7, 437 (1965).

66K Koltsov, Yu.I., Korovin, S.S. and Reznik, A.M., Zh. Neorg. Khim., 11, 2308 (1966).

66L Levin, V.I. and Kozlova, M.D., Radiokhimiya, 8, 533 (1966).

67B Biddle, P., Coe, A., McKay, H.A.C., Miles, J.H. and Watermann, M.J., J. Inorg. Nucl. Chem., 29, 2615 (1967).

68M Mitamura, R., Tokura, I., Nishimura, S., Kondo, Y. and Li, N.C., J. Inorg. Nucl. Chem., 30, 1019 (1968).

68S Startsev, V.N., Sannikov, Yu.I., Benyash, G.N. and Krylov, E.I., Zh. Neorg. Khim., 13, 1222 (1968).

IX. Solvent: Di(2-ethylhexyl) phosphoric acid ((H)DE-HP), HS

Reactions: $H^+ + S^- = HS$ (in aqueous solutions) (IX-1)

$$HS = \overline{HS} \qquad \text{(IX-2)}$$

$$2\,\overline{HS} = \overline{H_2S_2} \qquad \text{(IX-3)}$$

Although HS is used to a large extent in separations, both on the laboratory and the industrial scales, its fundamental equilibria have not been very extensively studied. Except for (H)DBP, however, which, because of its higher aqueous solubility, is less generally useful, but because of its role in the decomposition of TBP has been very thoroughly studied, HS has been better studied than other acidic organophosphorus reagents.

The aqueous solubility of HS is pH-dependent, through the common ion effect. A recent determination (69Y), giving the value 4.35×10^{-4} M in water at room temperature, is contradicted by reports of higher aqueous concentrations, namely 4.8×10^{-3} M in 0.2 M $HClO_4$ at 25° (in contact with a 1.10 M solution in CCl_4) (66S). Because of the possibility of acid impurities in these low concentration solutions of unspecified pH, these figures are all unreliable.

The acid dissociation in aqueous solution of HS is difficult to study because of the low solubility. In aqueous ethanol, titration methods led to log $Q_{(1)}$ = 3.35 in 98% (wt.) ethanol (65D) and log $Q_{(1)}$ = 3.2 in 70% (wt., or 75% vol.) ethanol (55B). In aqueous solution, the value log $Q_{(1)}$ = 2.85±0.05 at 25° has been reported to result from potentiometric titration and to have been confirmed by less accurate distribution data (66S). This value seems to be quite high, compared with other dialkyl phosphates which should all have similar acidities, and is high also in comparison with the data in aqueous ethanolic solutions. More reasonable values have been obtained from distribution data (with ^{32}P labeled HS) between 0.1 M (Na, H)ClO_4 aqueous media and decane (69S) or octane (70U).

Unfortunately, the values reported are in disagreement, being log $Q_{(1)}$ = 1.72 (at 25°) in the former work and 1.30 in the latter. Purification from acid impurities and measurements over wide pH and temperature ranges have been stressed in the earlier work, and its value for log $Q_{(1)}$ may be the more reliable. A very recently reported value (71L) log $Q_{(1)}$ = 0.15 in 1.0 M (Na, H)ClO_4 or (Na, H)HSO_4, from distribution between the aqueous phase and toluene, seems to be much too low. However, the dimerization in the aqueous phase has also been taken into account, and at the low concentrations encountered, this constant, log K (2 HS = H_2S_2) = 12.90 is extremely high, and has to be so to have any effect on the acid dissociation. Dimerization in the aqueous phase may not occur through the hydrogen bonding between the two =P(O)OH groups but rather, as in soap anions, by water-structure-enforced association of the alkyl chains. The dimerization constant would then be expected to be rather lower, and since this dimerization is coupled with the acid dissociation to give the observed distribution of HS, this may be a source of error and explain the unreasonably low value.

The equilibrium constants of reactions (2) and (3) are connected, and are obtained from distribution measurements for HS between an organic solvent and an appropriate aqueous phase -- such as 0.1 M $HClO_4$. The effect of aqueous phase composition on the distribution constants through salting out is small, as long as moderate concentrations of nonextractable electrolytes are present, and for lack of data for HS those for (H)DBP can be used:

$$\Delta \log Q_{(2)} = \log Q_{(2)}(1.0 \text{ M medium}) - \log Q_{(2)}(0.1 \text{ M medium}) = 0.14 \pm 0.06.$$

Data for 0.2 M $HClO_4$ should be within experimental error indistinguishable from those for 0.1 M (Na, H)ClO_4. Unfortunately, there are gross discrepancies between equilibrium quotients $Q_{(2)}$ and $Q_{(3)}$ reported by different authors for several solvents. Low values of the quotients may be due (69Y) to the presence of more water-soluble impurities, such as the monoalkyl ester. This may have been the case in a paper quoted earlier (66S) giving a much too high value for $Q_{(1)}$.

Similar low results for hydrocarbon and chlorinated solvents (68L) may also be due to impurities -- the details of this study have not been published, and a very recent paper by the same author (71L) gives, again without details, a much higher value for log $Q_{(2)}$ for toluene (4.79 as compared with his earlier value of 0.69). On the whole, the higher values seem to be more reliable, although they too have been criticized (71K). A similar situation exists with regard to $Q_{(3)}$, and earlier, low values (66S, 68L) should probably be discounted in view of later higher values (69Y, 70U and 71L) although the latter by no means agree among themselves. The work of Yagodin and Tarasov (69Y) gives only the combination log $Q_{(1)}{}^2Q_{(2)}{}^2Q_{(3)}$, but using log $Q_{(1)} = 1.30$ (70U) leads to values of log $Q_{(2)}{}^2Q_{(3)}$ for chloroform (13.38) and n-octane (11.43) in reasonable agreement with those in (70U) ($CHCl_3$ 13.92, C_8H_{18} 11.49). The new value (71L) for toluene is, however, considerably higher than that for benzene (14.68 vs. 12.63 (69Y)), much more than the difference for (H)DBP, the difference being only partly explained by accounting for $\Delta \log Q_{(2)}$. Thus, the situation with regard to $Q_{(2)}$ and $Q_{(3)}$ is seen to be completely unsatisfactory, and no recommended values for these quotients can at this time be presented.

References:

55B Blake, C.A., Brown, K.B., Coleman, C.F., White, J.C. and Schmitt, J.M., USAEC Report ORNL-1903 (1955).

65D Deptuła, C., and Minc, S., Nukleonika, 10, 421 (1965).

66S Szabo, E. and Szabon, J., Acta Chim. Acad. Sci. Hung., 48, 299 (1966).

68L Liem, D.H., in "Progress in Coordination Chemistry," (ed. M. Cais), Elsevier, Amsterdam, 1968, p. 140.

69Y Yagodin, G.A. and Tarasov, V.V., Radiokhimiya, 11, 148 (1969).

70U Ulyanov, V.S. and Sviridova, R.A., Radiokhimiya, 12, 47 (1970).

71K Kolarik, Z., Solv. Extr. Rev., 1, 1 (1971).

71L Liem, D.H. and Sinegribova, O., Acta Chem. Scand., 25, 277 (1971).

X. Solvents: Di(2-ethylhexyl)phosphoric acid ((H)DE-HP) , HS

Tri-n-butyl phosphate (TBP) , S'

Tri-n-octyl phosphine oxide (TO'PO) , S'

Reactions:

$$1/2\ \overline{H_2S_2} + \overline{S'} = \overline{HS \cdot S'} \quad (X\text{-}1)$$

$$\overline{H_2S_2} + \overline{S'} = \overline{H_2S_2S'} \quad (X\text{-}2)$$

The reaction of HS with S' should be quite similar to that of (H)DBP with S', since the alkyl chains should play a much lesser role in this direct interaction between the =P(O)OH and ≡P(O) groups, than in distribution or even dimerization reactions. Furthermore, the interaction of HS with S' = TO'PO is expected to be rather stronger than with S' = TBP, since the more basic oxygen of the former $\equiv P(O)$ group is expected to be a stronger hydrogen-bond acceptor than of the latter.

For S' = TO'PO, an infrared spectroscopic study (62B) in n-octane led to $\log Q_{(1)} = 1.40 \pm 0.08$ and this value was consistent with isopiestic vapor pressure data on the system HS-S'-octane. Based on an earlier report of this study, the value 1.48 ± 0.08 has been selected for use in synergistic distribution studies for uranium (59B, 63I) in octane or kerosene. Reaction (X-1) is accompanied always by reaction (X-2), but the constant for this is reported with much lower accuracy, $\log Q_{(2)}$ ranging from 1.7 to 2.6, with 2.3 ± 0.3 being a reasonable estimate, consistent with infrared, isopiestic and uranium extraction data in octane (59B, 62B), kerosene (63I) and cyclohexane (63Z).

For S' = TBP, there are a set of values (68L, 68La, 70U) which seem unreasonably high, in view of the TO'PO data, and also presume the occurrence of the reaction

$$\overline{HS} + 2\overline{S'} = \overline{HS \cdot S'_2} \qquad \text{(X-3)}$$

which has not been confirmed by others and is unlikely in view of the single hydrogen-bond donating ability of HS, though it might occur through polar interactions at the highest concentrations of S' used, ca. 1 M (68La). If these values are discounted, values very like those encountered in the analogous (H)DBP system are found, mainly from distribution data of HS in the presence of S' (66S, 69Y) and from synergistic extraction data for uranium (63B, 66Z) which, however, are not very accurate. A selection of values is shown in Table X-1 for several solvents, but their reliability cannot be said to have been established at all well.

Table X-1

Equilibrium Quotients for $\overline{(H)DE\text{-}HP} + \overline{TBP} = \overline{adduct}$

Diluent	log $Q_{(1)}$	log $Q_{(2)}$	Refs.
Octane, kerosene	0.0±0.1	0.7±0.1	63B,69Y
Hexane	-0.2±0.1	0.4±0.1	63B
Carbon tetrachloride	-0.5±0.1	0.1±0.1 (?)	63B
Benzene	-0.6±0.1	0.0±0.1	63B,66S

References:

59B Blake, C.A., Horner, D.E. and Schmitt, J.M., USAEC Report ORNL-2259 (1959).

62B Baker, H.T. and Baes, C.F., Jr., J. Inorg. Nucl. Chem., 24, 1277 (1962).

63B Baes, C.F., Jr., Nucl. Sci. Eng., 16, 405 (1963).

63I Ihle, H., Michael, H. and Murrenhof, A., J. Inorg. Nucl. Chem., 25, 734 (1963).

63Z Zangen, M., J. Inorg. Nucl. Chem., 25, 581 (1963).

66S Szabo, E. and Szabon, J., Acta Chim. Acad. Sci. Hung., 48, 299 (1966).

68L Liem, D.H., Acta Chem. Scand., 22, 753 (1968).

68La Liem, D.H., in "Progress in Coordination Chemistry," (M. Cais, ed.), Elsevier, Amsterdam, 1968, p. 740.

69Y Yagodin, G.A. and Tarasov, V.V., Radiokhimiya, 11, 148 (1969).

70U Ul'yanov, V.S. and Sviridova, R.A., Radiokhimiya, 12, 41 (1970).

XI. Solvent: Tri-n-butyl phosphate (TBP), S

Ligand: Nitrate (NO_3^-), A^-; Metal: Uranium (VI) (UO_2^{2+}), M^{2+}

Reaction: $M^{2+} + 2A^- + 2\overline{S} = \overline{MA_2S_2}$ (XI-1)

The extraction of uranium (VI) from nitrate solutions with TBP has received a very large amount of attention, and specifically, reaction (XI-1) has been examined quantitatively by many authors. Since Moore (49M) has originally proposed this formulation for the predominant extraction reaction in this system, this has been confirmed countless times over a very large concentration range. This reaction represents the predominating one, from very dilute S in many diluents: aliphatic hydrocarbons, such as n-hexane, iso-octane, n-decane, n-dodecane, kerosene of a variety of compositions; aromatic hydrocarbons, such as benzene and toluene; chlorinated hydrocarbons, such as carbon tetrachloride, chloroform, 1,2 dichloroethane; etc., up to concentrated solutions and even neat, undiluted (but, of course, water saturated) S. This holds true both for pure aqueous MA_2 solutions almost up to saturation, and for solutions containing HA or other nitrates besides MA_2, or even at trace concentrations of MA_2, in up to perhaps 13 M HA. In the latter cases, competition by the coextraction of HA (Section VII) must be taken into account. The following discussion contains detailed documentation over these statements. The few studies that report other reactions to occur in this system* do not deny, however, that reaction

*(57K) and (57R) suggest that

$$M^{2+} + 3A^- + H^+ + \overline{S} = \overline{MA_2 \cdot HA \cdot S} \quad \text{(XI-2)}$$

is important at very high HA concentrations; (67D) prefers to write $M(H_2O)_6^{2+}$ for M^{2+}, and therefore has $6H_2O$ released on the right hand side of XI-1, without, however, proving that this does occur; (61J) considers the ionic dissociation and disproportionation of MA_2 in the organic phase to MA^+ (solvated), $M_2A_3^+$ (solvated) and MA_3^- (solvated) in undiluted S; (63Rb) suggests that MA_2S_2 associates with the diluent, especially chloroform, to give solvates.

(XI-1) is the main one, over ordinary ranges of concentration, and particularly in dilute solutions, and these reports, therefore, need not be further considered.

For practical purposes, it is convenient to use the molar concentration scale in discussing reaction (XI-1), and therefore activities and activity coefficients, unless otherwise noted, will be in this scale, with pure water (the solubility of S in water is so low (Section V) as to make no difference) or infinitely dilute solutions of MA_2 and HA as the standard state for the aqueous phase, and infinitely dilute $\overline{S}$ and $\overline{MA_2S_2}$ in the (water saturated) diluent as the standard state for the organic phase. In the case of the neat extractant, i.e. undiluted S, infinitely dilute $\overline{MA_2S_2}$ in water-saturated S, of concentration 3.43±0.02 M at 25° (Section VI) will be the standard state.

The equilibrium constant for reaction (XI-1) is

$$K_{(1)} = (\overline{MA_2S_2})/(M^{2+})(A^-)^2(\overline{S})^2 = [\overline{MA_2S_2}]\overline{y}_M/[M^{2+}][A^-]^2y^3_{\pm MA_2}[\overline{S}]^2\overline{y}_S{}^2 =$$

$$= Q_{(1)}\overline{y}_M\overline{y}_S{}^{-2}y^{-3}_{\pm MA_2} \qquad \text{(XI-3)}$$

where $\overline{y}_M$ is the activity coefficient of $\overline{MA_2S_2}$ and $y_{\pm MA_2}$ is the mean ionic activity coefficeint of the component MA_2, whether or not HA or another nitrate is present in addition to MA_2, and $Q_{(1)}$ is the concentration quotient. The quantities $[\overline{MA_2S_2}]$ and $[M^{2+}]$ (or their ratio $D_M = \overline{C}_M/C_M$, provided $\overline{MA_2S_2}$ and M^{2+} are the representative species of M in the two phases) are obtainable from the analytical data. It is doubtful whether the corrections from C_M to $[M^{2+}]$, taking into account nitrate complexation and hydrolysis in the aqueous phase (e.g. 56Fb, 58C, 61M) are very meaningful in dilute solutions, and the magnitude of the complexation (to MA^+ or MA_2 aq.) and hydrolysis constants are such that systematic errors are not introduced by neglecting these corrections.* In the absence of added HA or another nitrate, i.e. in aqueous uranyl nitrate, $[A^-] = 2[M^{2+}]$ so that $[M^{2+}][A^-]^2$ is replaceable by $4[M^{2+}]^3$.

The quantity $[\overline{S}]$ is somewhat more difficult to estimate accuratley. The total analytical concentration of S at equilibrium, $\overline{C}_S$, differs somewhat from the initial value $\overline{C}_S°$ because of water coextration (or HA and water coextraction, when the former is present), causing volume changes, approximately proportional to $\overline{C}_S$. This, and of course the binding of S to extracted MA_2 and HA, must be taken into account. The formulation $[\overline{S}] = \overline{C}_S°(\overline{V}/\overline{V}°) - 2\overline{C}_M - \overline{C}_{HA}$, where $\overline{V}°$ is the initial, $\overline{V}$ the equilibrium volume of the organic phase, and $\overline{C}_S° = 3.655\times$(vol.% S in the diluent)/100 (62Da), is a good approximation, although $-\overline{C}_{HA}$ does not express adequately the correction for the concentration of $\overline{S}$ bound to $\overline{HA}$ over a very wide concentration range (see Section VII), and alternatives, such as $[\overline{S}] = \overline{C}_S/(1 + 0.16\ C_{HA})$ (58H, 61M) have been proposed. In the absence of added nitrate or HA, however, $[\overline{S}] = \overline{C}_S°(\overline{V}/\overline{V}°) - 2\overline{C}_M$ can be taken as essentially correct. A further point that needs consideration is whether to include in $[\overline{S}]$ a correction for the degree of hydration of S (64P), or whether to let S represent both hydrated and unhydrated S (Section VI). Since the constants will be evaluated for the limit of infinite dilution, i.e. constant water activity, this distinction is immaterial, and the usual convention, that S stands for all species of S, whether or not bound to components of constant activity (water and the diluent), will be used.

---

* In fact, the activity coefficient $\gamma_{\pm MA_2}$ has these corrections implicitly, and they should not be used with the experimental $\gamma_\pm$ values, but rather with virtual $\gamma'_\pm$ values, representing MA_2 as a strong electrolyte (58H).

Now that the concentration terms that make up $Q_{(1)}$ in (XI-3) have been specified, it is necessary to see how the activity coefficient terms can be evaluated. Most authors try to include $y_{\pm MA_2}$ in their calculations, and specify an effective constant

$$Q'_{(1)} = Q_{(1)}/y^3_{\pm MA_2} = K_{(1)}\bar{y}_M/\bar{y}_S^{\,2} \qquad \text{(XI-4)}$$

For aqueous MA_2 solutions, without added HA or another nitrate, this can be done unreproachably, by using the activity coefficient data of Robinson and Kim (51R) and density data of Kapustinskii and Lipilina (56K) to yield MA_2 activities. For solutions containing both MA_2 and HA, there exist recent vapor transpiration data (65D) that lead to the following equation

$$\ln a_M(m_H,m_M) = \ln a_M(m_M) + m_H[(1.73\pm0.08) - (0.34\pm0.09)m_M] \times$$
$$\times\ [1 - (0.080\pm0.002)m_H + (0.0020\pm0.0002)m_H^{\,2}] \qquad \text{(XI-5)}$$

where $a_M(m_M)$ is the activity (molal scale) of $MA_2 \cdot 6H_2O$ at a given fixed molality m_M, and $a_M(m_H,m_M)$ is that at the same molality but in the presence of m_H molal HA. This equation is valid in the range 0 - 2 m MA_2 and 1 - 13 m HA. This relationship, based on experimental evidence, has however <u>not</u> been used by the authors who studied reaction XI-1 in mixed MA_2-HA solutions, who have instead invoked the "ionic strength principle" or a vague "Harned's rule calculation." The former is based on data of Jenkins and McKay (54J) according to which (57R, 62Ra)

$$\log y_{\pm MA_2}(\mu = 3C_{MA_2} + C_{HA}) = \log y_{\pm MA_2}\ (\text{same } \mu,\ MA_2 \text{ only}) \qquad \text{(XI-6)}$$

to "sufficient accuracy," which seems to be at least ca. ±0.03 in $\log y_\pm$, and since $y_\pm$ enters in $Q'_{(1)}$ to the third power, this leads to an uncertainty of at least ±0.09 in $\log Q'_{(1)}$ for the cases of joint MA_2 and HA extraction. Therefore, data obtained in the absence of HA should be preferable for evaluating $Q'_{(1)}$ over those obtained in its presence.

Although the concentration quotient $Q_{(1)}$ cannot be expected to remain independent of the reactant concentrations over a wide range, it so happens that over the range 0.05 to 0.60 M MA_2 $\gamma_{\pm MA_2}$ is approximately constant, equalling 0.54 ± 0.03 (there exists a shallow minimum at ca. 0.27 M). If the activity coefficients in the organic phase vary in a manner that leaves $\bar{y}_M \bar{y}_S^{-2}$ essentially constant, then, within the above range of aqueous MA_2 solutions, log $Q_{(1)}$ is constant within ± 0.07, and $\log Q'_{(1)} - \log Q_{(1)} = 0.81$. This is useful when considering some of the earlier works, which concentrated on the practical aspects of the extraction and did not bother with the activity coefficients.

The activity coefficient ratio $\bar{y}_M \bar{y}_S^{-2}$ may, again, be expected to vary with the concentration of the solutes in the organic phase. If $\bar{C}_S$ is extrapolated to zero, $\bar{C}_M$ will also go to zero, since MA_2 will be less and less extracted (according to the second power of [S]), hence the standard state will be approached. Thus

$$\log K^{\infty}_{(1)} = \lim (\bar{C}_S \rightarrow 0) \log Q'_{(1)} = \log K_{(1)} \qquad \text{(XI-7)}$$

since $\lim \bar{y}_M = \lim \bar{y}_S = 1$. Another approach is to evaluate $\bar{y}_M$ and $\bar{y}_S$ separately, perhaps in the binary solutions: S-diluent and MA_2S_2-diluent, say from vapor pressure measurements (63P,67D, 67R), with or without an aqueous phase present. The validity of these estimates will be discussed individually below.

In the following, several diluents, for which there exist comparative studies, and neat S, will be discussed in detail.

XI-A. Benzene diluent

The studies (64Na and 67R) contain data from which constants have been calculated, while (64R) and (60D) cite other studies (59Hb, and 60H and 58Ha, respectively) as the source of the constants they report. Furthermore, (63Pa) and (63Rb) cite constants, without giving the source of the data on which they are based. The temperature was not stated in these studies, but room temperature, 20-25°, is understood in all of them.

The constant log $Q'_{(1)}$ = 2.28 given in (64R/59Hb) is based on data obtained at low MA_2 concentrations (3×10^{-4} to 3×10^{-3} M) in 0.5 M HA, so that the activity coefficient $\gamma_{\pm MA_2}$ is not well established. The other studies used pure aqueous MA_2 solutions, and a range of $\overline{C}_S$ values. If log $Q'_{(1)}$ is plotted against $\overline{C}_S$ a rather smooth curve results (Fig. XI-1a), which extrapolates to log $K^{\infty}_{(1)}$ = lim ($\overline{C}_S \to 0$) log $Q'_{(1)}$ = 2.20±0.10. The value at 10% vol. S, $\overline{C}_S$ = 0.35 M, is log $Q'_{(1)}$ = 2.07±0.05, essentially the value given, without discussion, by Rozen (63Rb), and may be compared with the values for other diluents in Table XI-1.

XI-B. Chloroform diluent

Chloroform has a hydrogen atom capable of hydrogen-bonding with the phosphoryl oxygen of TBP, hence is expected to lower the ability of the latter to extract MA_2. It should be remembered that although $\overline{y}_S$ extrapolates to unity at the limit of $\overline{C}_S \to 0$, this is the value relative to chloroform as the standard state. It is the constant $K^{\infty}_{(1)}$, rather than the activity coefficient, which expresses the free energy change from pure S as the standard state to the infinitely dilute solution standard state.

Several studies examined reaction XI-1 in chloroform, (60Pa) and (62Pa) giving data for 20°, (64Na) and (67R) for an unspecified temperature (i.e. room temperature, 20-25°), while (63P) (63Rb) and (64P) cite only constants without giving their source. The latter study reported only on mixed chloroform-carbon tetrachloride solutions, which cannot be extrapolated to pure chloroform, hence cannot be used in the present context.

The data can be extrapolated reasonably well (Fig. XI-1b) to $\overline{C}_S$ = 0, yielding log $Q'^{\infty}_{(1)}$ = -0.25±0.10 and log $Q'_{(1)}$= -0.17±0.05 for $\overline{C}_S$ = 0.35 M (10% vol. S in chloroform).

The data of (63P) include some for $\overline{C}_S$ = 1.1 M, obtained for 0.02 - 0.07 M MA_2 in 0.5 - 4.0 M HA, where it was assumed that $y_{\pm MA_2}$ is the same as in pure aqueous MA_2 solutions at the same μ as in the mixed MA_2-HA solutions. In spite of the clear dependence of log $Q'_{(1)}$ on $\overline{C}_S$, the same constant is given as at $\overline{C}_S$ = 1.47 M in MA_2 solutions. Far from confirming the applicability of the "ionic strength principle" for the mixed MA_2-HA solutions, this equality shows it to be invalid, since a definitely lower value of log $Q'_{(1)}$ is expected at the lower $\overline{C}_S$. The value of log $Q'_{(1)}$ given in (63Rb) is definitely too high for $\overline{C}_S$ = 0.35 M, but since no data are given and the value of log $Q'_{(1)}$ is only quoted in a footnote, it must be discounted. The values given in the other studies quoted conform, within their stated limits of error, to the data given, and to one another, considering the variation of $\overline{C}_S$ (Fig. XI-1b).

XI-C. Carbon tetrachloride diluent.

There are many studies concerned with reaction XI-1 in carbon tetrachloride, and most of them lead to $Q'_{(1)}$ values highly consistent with each other. The most thorough study seems to be that of Aartsen and Korvezee (64Aa), at least they present the largest amount of data, with the best internal consistency. At the first glance it seems peculiar that their reported log $Q'_{(1)}$ = 1.68±0.01 should be independent of $\overline{C}_S$ at the high concentrations used by these authors, 50-100% vol. S in CCl_4, corresponding to 1.74 to 3.43 M, or to mole fractions of $\overline{X}_S$ = 0.46 to 1.00. Even more peculiar is that it should equal, within the limits of error, the value of log $K^{\infty}_{(1)}$ = 1.70±0.03 obtained by extrapolation to very low S concentrations (see below). However, since log $K^{o}_{(1)}$ = 1.74±0.04

(XI-E below), and there is only a very mild dependence of $\overline{y}_M/\overline{y}_S^2$ on $\overline{C}_S$ in this case*(63P, 67R), the near independence of log $Q'_{(1)}$ from $\overline{C}_S$ should not be astonishing.

Other studies, somewhat less well documented with data (62Pa, 63P, 63Pa, 63Pa/58B, 63Rb, 64Na, 64P, 64R, 64R/58B, 67R) all yield log $Q'_{(1)}$ values in the range 1.63 to 1.77, and weighting the reported values in accordance with the quality of their data (where given) or the limits of error given by the author, gave the average value 1.70±0.03, independent of $\overline{C}_S$ and $\overline{C}_M$ (for C_M < 0.7 M), and of the temperature in the interval 20-25°C. Although within this range of temperatures, it is impossible to give a definite trend because of the imprecision of the data, one may still discern a trend among those studies that report the temperature of measurement. Those that report 20° (62Pa) tend to cluster around higher values, log $Q'_{(1)}$ = 1.72 to 1.75 (±0.03), while those that report 25° (64Aa, 64R) tend to cluster around lower values, log $Q'_{(1)}$ = 1.63 to 1.68 (±0.04). Again, while not statistically significant per se, this is in line with the temperature dependence, studied over a wider range (65R,

---

*Even for carbon tetrachloride do the individual values of $\overline{y}_M$ and $\overline{y}_S$ vary considerably with $\overline{C}_S$, as much perhaps as for the other diluents considered here. It is mainly by accident that the ratio $\overline{y}_M/\overline{y}_S^2$ in the present case is nearly independent of $\overline{C}_S$.

71Ka). In all these studies, pure aqueous solutions of MA_2 were employed, and thus directly comparable values of $Q'_{(1)}$ were obtained.

In addition to these studies, there exist those in which mixed MA_2 and HA solutions were used. Some of the older ones gave only the concentration quotient $Q_{(1)}$ (49M: $\log Q_{(1)} = 0.89 \pm 0.04$, and 59V: $\log Q_{(1)} = 0.96 \pm 0.03$), and these can be converted as discussed above to $\log Q'_{(1)}$ by adding 0.81 ± 0.07 to give 1.70 ± 0.08 and 1.77 ± 0.08 respectively. The latter work did attempt an extrapolation to zero concentrations of added nitrate (at trace MA_2), but the value obtained, $\log Q^{\infty}_{(1)} = 1.10$ is very far from those calculated at higher concentrations with the incorporation of valid corrections for $\gamma_{\pm MA_2}$. Another attempt (68A) to consider the competitive extraction of MA_2 and HA derived the equilibrium constants from the maximum in the distribution curve, with constant $\overline{C}_S$, low C_M and varying C_{HA}. However, the value $\log Q'_{(1)} = 1.23$ found, as well as the "thermodynamic value of 1.34" cited without further explanation, are invalidated because of the wrong assumptions $\overline{y}_{HAS} \approx 1$ and $\overline{y}_S = 1$ (the latter implicitly made over the concentration range where $[\overline{S}]$ varies from zero to 0.7 M in CCl_4), among other errors. These very low values for $\log Q'_{(1)}$, and similarly unexplained low results that were obtained in (60N) and (61N), the values of $\log Q_{(?)} = 1.43$, and $\log Q_{(?)} = 1.29$ respectively, where, however, the exact nature of the "constant" cited was not specified, cannot be further discussed.

In other studies, however, $\gamma_{\pm MA_2}$ has been evaluated according to the "ionic strength principle" (XI-6) and yield consistent data (63P, 63Pa/60Pa, 65R, 65Sd) in the range $1.63 < \log Q'_{(1)} < 1.90$. Evaluated as above by proper weighting, these give an average of 1.74 ± 0.04, which is seen to be somewhat, statistically insignificantly, higher than the value 1.70 ± 0.03 found in pure aqueous MA_2 solutions. The conclusion from this comparison is that the "ionic strength principle" gives a fair estimate for $\gamma_{\pm MA_2}$ in the mixture, yielding a constant only slightly high (compare also the case of chloroform solutions above (XI-B)). In most of the mixed solutions, trace MA_2 has been used with up to 4 M HA, but in one (49M), 0.005 - 1.5 M MA_2 and 0-2.5 M HA gave values of log Q internally consistent to within ±0.04 in spite of the very wide concentration ranges employed.

Attempts have been made to evaluate $\bar{y}_M$ and $\bar{y}_S$ for this case, in order to be able to calculate $K_{(1)}$. The most detailed discussion has been presented by Rozen et al. (67R, 67Ra), while earlier discussions were given by Pushlenkov (63P) and by Rozen (63Rb) and their coworkers. In the earlier studies, it was assumed that mutual interactions of $\bar{S}$ and of $\overline{MA_2S_2}$ can be neglected.*

In the latest study (67R, 67Ra), however, the small mutual interaction of $\overline{MA_2S_2}$ and $\bar{S}$ was also taken into account, as well as the presence of water (cf. also 63P). An eight-constant equation

$$\ln \bar{y}_M\bar{y}_S^{-2}(0) = 2b_{SD} - b_{MD} + \phi_S{}^\circ b_{MS} + \phi_W{}^\circ b_{MW} + \phi_S{}^\circ\phi_W{}^\circ[\Delta_{MW} - c_{SMW} - (V_M/V_S)(b_{SW} - 2\phi_S{}^\circ\Delta_{SW})] \qquad \text{(XI-9)}$$

---

*(63P) based this assumption on the fact (data not shown) that "the average activity coefficients of the sum of the nonvolatile components at constant $\bar{X}_{\text{diluent}}$ ... vary approximately linearly from $\bar{f}_M$ to $\bar{f}_S$ as a function of the ratio of S to MA_2S_2 in the liquid phase" for the diluent chloroform. (63Rb) based the assumption that "molecules of both TBP and the solvate (MA_2S_2) react markedly more with C_6H_6, CCl_4 and $CHCl_3$ than with one another" on the fact that the thermal effects for the binary systems are negative (62Rc) and that "in the presence of only Van-der-Waals forces the reaction relation is the reverse ... and positive nonideality is observed."

where (0) denotes the absence of HA in the system, ϕ° is the volume fraction, discounting the diluent, b_{ij} and Δ_{ij} are the binary, and c_{ijk} a ternary interaction constant, V_i are molar volumes of the pure components (given as V_S = 270 ml and V_M = 606 ml), (67R), and M, W and D referring to $\overline{MA_2S_2}$, $\overline{H_2O}$ and the diluent respectively. The whole expression corrects from the standard state of the pure components to the standard state of the infinite dilution of M, S and W in the diluent. The binary constants b_{SD} and b_{MD} are obtained from solvent activity curves shown in the quoted study, and the mutual interaction constant b_{MS} from data published elsewhere (64Rb) for dry systems, while b_{MW}, Δ_{MW}, b_{SW} and Δ_{SW} are given approximate values in the paper, without any details (also an arbitrary value of $\phi_S{}^\circ = 1 - \phi_W{}^\circ = 0.94$ is assumed), and the ternary constant c_{SMW} is completely disregarded. The present reviewer cannot see how valid independent values of these interaction parameters, particularly those involving water, can be obtained, to any accuracy that justifies their inclusion -- in fact (67R) does not quote a value for b_{MS} -- so that only an approximate evaluation is possible. Unfortunately, (63P) did not specify how "considering the aqueous layer" was conducted in calculating their $\overline{f}_M$ and $\overline{f}_S$ (mole-fraction scale) to obtain $\log K_{(1)} = 7.70 \pm 0.05$ (X-scale) as obtained from the data presented, or $\log K_{(1)} = 7.61$ given in a table. This value cannot, therefore, be compared with the other values. Rozen et al. (67R), however, showed that the independently obtained b_{SD} and b_{MD} are sufficient for accounting for a major fraction of the diluent effect, and that at the limit of infinite dilution $f_S{}^\infty$(volume-fraction scale) = 0.42, $\overline{f}_M^{\infty}$ = 0.21 (relative to pure dry S and MA_2S_2 as the standard states). Therefore, the correction from log $K^{\infty}_{(1)}$ to log $K^{o}_{(1)}$ can be calculated (converting to $\overline{y}$, molar activity coefficients, $\overline{y}_i^{\infty} = \overline{f}_i^{\infty} V_i / V_{CCl_4}$):

$$\log K^{o}_{(1)} = \log K^{\infty}_{(1)} + \log \overline{y}^{\infty}_{M} - 2 \log \overline{y}^{\infty}_{S} =$$

$$= (1.70 \pm 0.03) + 0.116 - 2\times 0.070 = 1.68 \pm 0.03 \qquad \text{(XI-10)}$$

noting that according to the definition of the standard states, $\overline{y}_M^{\circ} = \overline{y}_S^{\circ} = 1$. However, in order to compare this value with practical results, the standard state must be redefined as the water-saturated neat S and MA_2S_2, in equilibrium with pure water. This point will be discussed below (XI-E).

XI-D. Paraffin hydrocarbon diluents

Since practical extraction processes according to reaction (XI-1) are usually conducted in kerosene as a diluent, it is not surprising that various paraffinic hydrocarbon diluents have been widely used also for more fundamental studies, leading to equilibrium constants. For a given, carefully controlled set of conditions, it is possible to discern a clear trend among various hydrocarbons, according to either chain length or molar volume: $Q'_{(1)}$ decreases with increasing molecular size of the diluent (67R). On the whole, however, on comparing results by different authors under widely varying conditions, this distinction is difficult to sustain, and hexane (63P, 63Rb, 67R) decane (71Ka), dodecane (59S, 67O) and kerosene of various descriptions (Amsco - 125-90W, "sat-

C*

urated hydrocarbons," etc., 56Aa, 56Fa, 57R/57K, 58H/57Aa, 60N, 62F, 62R, 67R) can be treated as one kind of diluent, called henceforth "kerosene."

"Kerosene," like benzene and chloroform, and in distinction from carbon tetrachloride, is a diluent for which the equilibrium quotients are quite sensitive to the extractant concentration (since $Q^{\infty}{}_{(1)}$ is quite different from $K^{o}_{(1)}$), hence a plot of log $Q'_{(1)}$ against $\overline{C}_S$ is required to obtain a consistent set of equilibrium quotients, which can be extrapolated to infinite dilution (see Fig. XI-d). As before, equilibrium quotients log $Q_{(1)}$ are converted to log $Q'_{(1)}$ for pure aqueous MA_2 solutions by adding 3 log $\gamma_{\pm}$ and for mixed MA_2-HA solutions by adding 3 log $\gamma_{\pm}$ (according to XI-6) ± 0.09. The data of several authors (56Aa, 57R/57K, 62R, 63P, 63Rb, 67R, 71Ka*) fall on a smooth curve. Those of several others do not conform, and the deviations can be explained as follows. In (58H/57Aa) trace concentrations of MA_2 in HA were used, and $y_{\pm MA_2}$ has been calculated in a manner differing from (XI-6), from the data of (54J). Similarly, (60N), (62F) and (67O) give $Q_{(1)}$ values, which cannot be converted unambiguously to $Q'_{(1)}$ values because of the nonavailability of $y_{\pm MA_2}$ data for the situations treated in these studies (mixed $NaNO_3$-HNO_3 and NH_4NO_3-HNO_3 solutions in the former two, and a μ-dependence in the latter, which partly incorporates these coefficients).

Mention should also be made of the work of Solovkin (64S) who, instead of using mean ionic activity coefficients $y_{\pm MA_2}$, for converting concentration quotients to equilibrium constants, used the reciprocal of "the equivalent surface

---

*In this study (71Ka), there is probably a misprint in the definition of $Q'_{(1)}$, which has 2 in the denominator rather than the required 4. If the constant has been calculated correctly, i.e. with a 4, then the value conforms to the curve. If the constant has been calculated with the formula as printed, i.e. with a 2, it must be corrected, and then it would fall considerably below the curve.

density of distribution of water molecules in the first coordination layer of the ions" (60Sb), that for M^{2+} being 0.063, that for A^- being 0.044, so that the resultant factor is $(0.063 \times 0.044^2)^{-1}$ or 8.2×10^3. Dividing the values of the "extraction constants (Cp)" given in this work by this factor yields the concentration quotients $Q_{(1)}$ given by the authors who are quoted in this work (57Aa, 57Bb, 58A, 58Ca, 59Sb, 60Se), without, however, yielding $Q'_{(1)}$ values comparable with the other data discussed here. (64R) quotes (59Ha) as the source for the data yielding the constant cited, $\log Q'_{(1)} = 2.11$ for dodecane diluent. However, (59Ha) does not mention at all dodecane (nor kerosene) as a diluent used in this work, so that the quotation of (59Ha) must be a misprint. For the other studies, the data have been weighted as described above for the purpose of the extrapolation. The extrapolation yields

$$\log K^{\infty}_{(1)} = \lim (\overline{C}_S \rightarrow 0) \log Q'_{(1)} = 2.25 \pm 0.08 \qquad \text{(XI-11)}$$

and at $\overline{C}_S = 0.35M$ (10% vol.)

$$\log Q'_{(1)} = 2.05 \pm 0.05. \qquad \text{(XI-12)}$$

As discussed in XI-C, attempts have been made to estimate $\overline{y}_M$ and $\overline{y}_S$ independently also for the case of "kerosene" diluents. Contrary to the cases of carbon tetrachloride, chloroform and benzene, the binary interactions of S and of MA_2S_2 with "kerosene" have positive deviations from ideality, i.e. signifying very weak interactions, permitting stronger interactions of each solute with itself, and also of the two solutes with each other, so that b_{MS} in (XI-9) should not be small in comparison with b_{SD} and b_{MD}. Application of (XI-10) should give the same value of $K^o_{(1)}$ as found above. However, with the parameters given for hexane (67R) $\overline{f}^{\infty}_S = 4.36$ and $\overline{f}^{\infty}_M = 12.3$, conversion to the molar concentration scale yields

$$\log K^{o}_{(1)} = 2.25 \pm 0.08 + 1.75 - 2 \times 0.96 = 2.08 \pm 0.08 \qquad \text{(XI-13)}$$

which differs considerably from the value given above (XI-10), obtained from data in carbon tetrachloride. Since at the limit of infinite dilution the mutual interaction term with b_{MS} should be zero ($\phi_S^{\circ} = 0$ in (XI-9)), the discrepancy may be due to inaccuracies in the quoted values of $\overline{f}_i^{\infty}$. The constant given by Pushlenkov (63P) on a mole fraction scale, $\log K^{o}_{(1)} = 7.18$, is again not comparable, even after conversion of the concentration scale, because the means for "considering the aqueous layer," which were employed in arriving at this value, were not disclosed.

XI-E. Undiluted, neat S

Undiluted S is inconvenient for the purpose of extraction because of its high viscosity and its density, which is too near that of the aqueous solutions for easy phase separation. Still, several studies have reported equilibrium constants for this case. In this case, although $y^{-3}_{\pm MA_2}$ corrects $Q_{(1)}$ to zero MA_2 concentration in the aqueous phase, the resulting $Q'_{(1)}$ still depends on the concentration of MA_2 in the organic phase, and extrapolation to $\overline{C}_M \rightarrow 0$ is necessary to arrive at $Q'^{\infty}_{(1)} = K^{o}_{(1)}$ values.

An early study (59Ha) reports the equilibrium constant 2230 (probably for 25°, though not stated explicitly) in a mixed concentration scale: molal for MA_2 and mole fraction for S, assuming $\overline{x}_S = 1$ for the undiluted extractant. However, the equilibrium constant given corresponds to $\overline{x}_S = 0.489$ (Section V), so that conversion to the molar scale, with $\overline{C}_S = 3.47$ M for water-saturated S, entails division by $(3.47/0.489)^2$, and conversion of $\overline{m}_{MA_2}$ to $\overline{C}_{MA_2}$ demands multiplication by the density of water-saturated S, $\rho_{S(W)} = 0.977$, so that

$$Q'^{\infty}_{(1)} = 2230 \times 0.977/(3.47/0.489)^2 = 44 \qquad \text{(XI-14)}$$

results, or $\log Q'^{\infty}_{(1)} = 1.64$. No estimate of the precision of this constant can be given, but it can be seen to be based on five experimental points falling on the linear portion of the curve in Fig. 1 of (59Ha), with deviations ≤ 0.02 log units.

A study conducted in the presence of high and varying concentrations of HA or of calcium nitrate (57Ba) cannot be employed for calculating $Q'_{(1)}$, but several others (57R/57K, 64Aa, 67R) which employed pure aqueous MA_2 (or solutions acidified to 0.05 M HA (57K)) can be used for an extrapolation of $\log Q'_{(1)}$ vs. $\overline{C}_M$, yielding

$$\log Q'^{\infty}_{(1)} = 1.72 \pm 0.04. \qquad \text{(XI-15)}$$

In addition to the value from (59Ha), 1.64, which is somewhat low, but probably within its precision consistent with the extrapolated value, there is also the value cited in (67R), $\log Q'^{\infty}_{(1)} = 1.65$ as a result of the non-linear extrapolation from four points, and $\log Q'_{(1)} = 1.63$ (64N), both on the low side. On the other hand, there is the detailed study of Davis and Mrochek (67D), which gives for 23-27° the value

$$\log Q''^{\infty}_{(1)} = \lim\,(\overline{C}_M/\overline{C}_S \rightarrow 0) \log\,[\overline{C}_M a_{H_2O}{}^6/4C_M{}^3\, y_{\pm MA_2}{}^3\, \overline{a}_S{}^2] = 3.425 \pm 0.010 \qquad \text{(XI-16)}$$

from a least-squares analysis of 27 data points. Since at the limit $a_{H_2O} = 1$ and $\overline{a}_S = 0.515$, conversion to the molar scale for S gives

$$\log Q'^{\infty}_{(1)} = 1.76 \pm 0.01, \qquad \text{(XI-17)}$$

somewhat on the high side of the extrapolated value from the other studies. Another high value is obtained from (68M), which analyzed the phase diagram of the MA_2-S-H_2O system given in (62A) with the aid of a computer, taking into account all the binary interactions, and showing that the ternary ones are negligible. The mixed-scale constant obtained (m-scale for aqueous MA_2, x-scale for the organic phase), on conversion to the C-scale by division by $\overline{C}_{S(hydr)} = 3.43$ M, gives log $K^{o}_{(1)} = 1.82 \pm 0.01$. Weighting the above results according to their internal precision, the average value would be log $K^{o}_{(1)}$ = log $Q'_{(1)} = 1.75 \pm 0.02$, which is a very precise value, notwithstanding the difficulties in the determination, and predominantly because of its freedom from assumptions of limited validity, this value is judged to be also highly accurate. It compares well with log $K^{o}_{(1)} = 1.68 \pm 0.03$ (XI-10) obtained from data for carbon tetrachloride, but poorly with the value log $K^{o}_{(1)} = 2.08 \pm 0.08$ obtained from those for hexane, as already discussed.

In order to convert $K^{o}_{(1)}$, based on the standard state of water-saturated S, to a constant based on pure, dry S as the standard state, the activity coefficients of both $\overline{S}$ and $\overline{MA_2S_2}$ are required, i.e. the standard Gibbs energies of transfer of both solutes from the one state to the other. While that for S is known (Section V), that for MA_2S_2 is not known definitely, so that this calculation cannot be performed.

Table XI-1

Equilibrium constants for the reaction $UO_2^{2+} + 2NO_3^- + 2\ \overline{TBP} = \overline{UO_2(NO_3)_2(TBP)_2}$ at 20-25°

Diluent	log $Q'^{\infty}_{(1)}$	log $Q'_{(1)}$ at $\overline{C}_S = 0.35$ M	log $K^{o}_{(1)}$
Benzene	2.20±0.10	2.07±0.05	
Chloroform	-0.25±0.10	-0.17±0.05	
Carbon tetrachloride	1.70±0.03	1.70±0.03	
"Kerosene"	2.25±0.08	2.05±0.05	
Undiluted	1.75±0.02		1.75±0.02

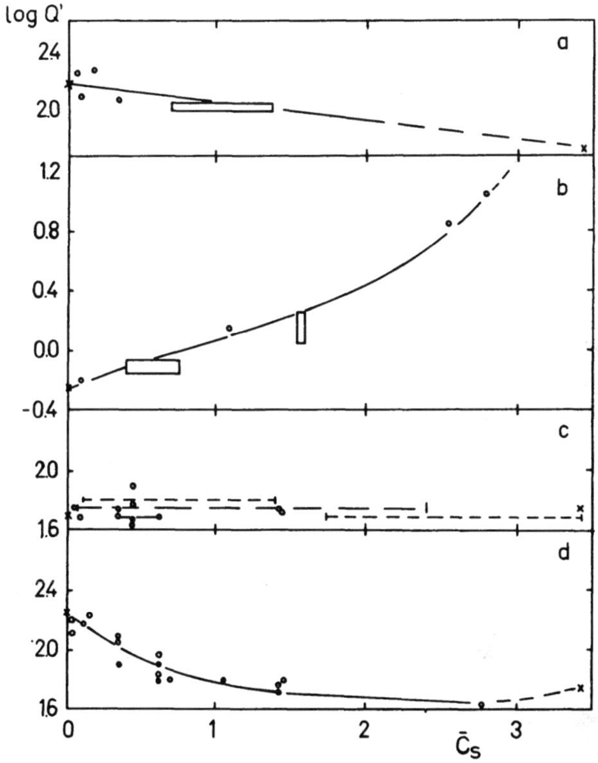

Fig. XI-1

Effective equilibrium constants, log $Q'_{(1)}$, plotted against the extractant concentration, $\overline{C}_S$, for a) benzene, b) chloroform, c) carbon tetrachloride, d) "kerosene." The constants are given the estimated precision limits appearing in the original papers, and values pertaining to a range of $\overline{C}_S$ values are so indicated.

References:

49M Moore, R.L., USAEC Report AECD 3196 (1949).

51R Robinson, R.A. and Lim, C.K., J. Chem. Soc., 1951, 1840.

54J Jenkins, J. and McKay, H.A.C., Trans. Faraday Soc., 50, 107 (1954).

56A Andelin, R.L., Anderson, E.L. and McVey, H.W., USAEC Report IDO-14381 (1956).

56Fa Flanary, J.R., Proc. Intl. Conf. Peaceful Uses At. Energy, 1st, Geneva, 1955, United Nations, 9, 528 (1965).

56Fb Fletcher, J.M., in "Progress in Nuclear Chemistry," Sec. III, Process Chemistry, Pergamon, Oxford, Vol. 1, 105 (1956).

56K Kapustinskii, A.F. and Lipilina, I.I., Izv. Akad. Nauk. SSSR, Otd. Khim. Nauk, 6, 661 (1956).

57A Alcock, K., Bedford, F.C., Hardwick, W.H. and McKay, H.A.C., J. Inorg. Nucl. Chem., 4, 100 (1957).

57Ba Bernstein, B. and Rydberg, J., Acta Chem. Scand., 11, 1173 (1957).

57Bb Best, G., McKay, H.A.C. and Woodgate, P.J., J. Inorg. Nucl. Chem., 4, 315 (1957).

57K Karpacheva, S.M., Khorkhorina, L.P. and Rozen, A.M., Zh. Neorg. Khim., 2, 1441 (1957).

57R Rozen, A.M., Zh. Neorg. Khim., 2, 1956 (1957).

58A Alcock, K., Best, G., Hesford, E. and McKay, H.A.C., J. Inorg. Nucl. Chem., 6, 328 (1958).

58B Burger, L.L., J. Phys. Chem., 62, 590 (1958).

58C Codding, J.W., Jr., USAEC Report IDO-14454 (1958).

58Ca Codding, J.W., Haas, W.O. and Hermann, F.K., Ind. Eng. Chem., 50, 145 (1958).

58H Hesford, E. and McKay, H.A.C., Trans. Faraday,Soc., 54, 573 (1958).

58Ha Healy, T.V. and Kennedy, J., British Report AERE C/R 2686 (1958).

59Ha Healy, T.V., Kennedy, J. and Waind, G.M., J. Inorg. Nucl. Chem., 10, 137 (1959).

59Hb Healy, T.V. and Kennedy, J., J. Inorg. Nucl. Chem., 10, 128 (1959).

59S Siddall, T.H., III., J. Am. Chem. Soc., 81, 4176 (1959).

59Sa Shevchenko, B.V., Solovkin, A.S. et al., Radiokhimiya, 1, 257 (1959).

59V Voden, V.G., Nikitina, G.P. and Pushlenkov, M.F., Radiokhimiya, 1, 121 (1959).

60D Dyrssen, D. and Kuca, L., Acta Chem. Scand., 14, 1945 (1960).

60H Hardy, C.J., British Report AERE-R 3267 (1960).

60N Naito, K., Bull. Chem. Soc. Japan, 33, 363 (1960).

60P Petrov, K.A., Shevchenko, V.B. et al., Zh. Neorg. Khim., 5, 498 (1960).

60S Solovkin, A.S., Zh. Neorg. Khim., 5, 2119 (1960).

60Sa Sato, T., J. Inorg. Nucl. Chem., 16, 156 (1960).

61J Jezowska-Trzebiatowska, B. and Chmielowska, M., J. Inorg. Nucl. Chem., 20, 106 (1961).

61M Marcus, Y., J. Phys. Chem., 65, 1647 (1961).

61N Nomura, S. and Hara, R., Anal. Chim. Acta, 25, 212 (1961).

62A Aartsen, J.J. van, Ph.D. Thesis, Groningen, 1962.

62D Davis, W., Jr., Nucl. Sci. Eng., 14, 159 (1962).

62F Fomin, V.V., Kartushova, R.E. and Maiorova, E.P., in "Extraktsiya", Atomizdat, Moscow, Vol. 2, 37 (1962).

62P Pushlenkov, M.F., Komarov, E.V. and Shuvalov, O.N., Radiokhimiya, 4, 543 (1962).

62R Rozen, A.M. and Moiseenko, E.I., in "Extraktsiya", Atomizdat, Moscow, Vol. 2, 235 (1962).

62Ra Rozen, A.M., Khorkhorina, L.P., Karpacheva, S.M. and Agashkina, G.D., Radiokhimiya, 4, 591 (1962).

63P Pushlenkov, M.F. and Shuvalov, O.N., Radiokhimiya, 5, 543 (1963).

63Pa Pushlenkov, M.F. and Shuvalov, O.N., Radiokhimiya, 5, 551 (1963).

63R Rozen, A.M., Khorkhorina, L.P., Yurkin, V.G. and Novikova, N.M., Dokl. Akad. Nauk SSSR, 153, 1387 (1963).

64A Aartsen, J.J.V., and Korvezee, A.E., Trans. Faraday Soc., 60, 510 (1964).

64N Nikolaev, A.V., Torgov, V.G., Mikhailov, V.A. and Kotlyarevskii, I.L., Dokl. Akad. Nauk SSSR, 156, 616 (1964).

64P Pushlenkov, M.F. and Komarov, E.V., Radiokhimiya, 6, 426 (1964).

64R Rozen, A.M. and Nikolotova, Z.I., Zh. Neorg. Khim., 9, 1725 (1964).

64Ra Rozen, A.M., Khorkhorina, L.P., Yurkin, V.G. and Novikova, N.M., Proc. Intl. Conf. Peaceful Uses Atomic Energy, 3rd, Geneva, 1964, United Nations, 10, 338 (1964).

64S Solovkin, A.S., Zh. Neorg. Khim., 9, 746 (1964).

65D Davis, W.,Jr., Lawson, P.S., DeBruin, H.J., and Mrochek, J., J. Phys. Chem., 69, 1904 (1965).

65R Rozen, A.M., Nikolotova, Z.I., Petrov, K.A., Skotnikov, A.S. and Teterin, E.G., Radiokhimiya, 7, 517 (1965).

65S Shuvalov, O.N., Radiokhimiya, 7, 265 (1965).

67D Davis, W. Jr. and Mrochek, J., in "Solvent Extraction Chemistry," (D. Dyrssen, J.-O. Liljenzin, and J. Rydberg, Eds.), North-Holland, Amsterdam, 1967, p. 283.

67O Ochsenfeld, W., Schneider, H. and Baumgärtel, G., German Report KFK-671 (1967).

67R Rozen, A.M., Khorkhorina, L.P. and Yurkin, V.G., Zh. Neorg. Khim., 12, 244 (1967).

67Ra Rozen, A.M., in "Solvent Extraction Chemistry" (D. Dyrssen, J.-O. Liljenzin and J. Rydberg, Eds.), North-Holland, Amsterdam, 1967, p. 195.

68A Asano, M., Tech. Report Engng. Res. Inst. Kyoto Univ., No. 138 (1968).

68M Mikhailov, V.A., Ponomareva, L.I. and Nazin, A.G., Zh. Fiz. Khim., 42, 318 (1968).

71K Komarov, E.V., Shpunt, L.B. and Pushlenkov, M.F., Radiokhimiya, 13, 380 (1971).

XII. Solvent: Tri-n-butyl phosphate (TBP) , S

Ligand: Nitrate (NO_3^-), A^-; Metal: Plutonium (IV), (Pu^{4+}), M^{4+}

Reaction: $M^{4+} + 4A^- + 2\bar{S} = \overline{MA_4S_2}$ (XII-1)

The extraction of plutonium (IV) from nitrate solutions with TBP is an important step in the technology of nuclear fuel reprocessing, and has therefore been studied fairly thoroughly. Because of the hazards involving plutonium, most workers avoided handling macro quantities, and conducted their investigations with tracer concentrations of plutonium, in the presence of macro concentrations of nitric acid and possibly also of uranium (VI). This circumstance makes the estimation of the activity coefficients much more difficult than in the case of uranium extraction (Section XI). A further source of uncertainty is the much higher importance of the activity coefficient term, since $y_{\pm MA_4}$ enters the expressions for the equilibrium constant in the fifth power.

There is general agreement that the stoichiometry of the extraction reaction is given by (XII-1) (57B, 58H). The equilibrium constant of this reaction is

$$K_{(1)} = (\overline{MA_4S_2})/(M^{4+})(A^-)^4(\bar{S})^2 = [\overline{MA_4S_2}]\bar{y}_M/[M^{4+}][A^-]^4 y_{\pm MA_4}{}^5[S]^2\bar{y}_S{}^2 =$$

$$= Q_{(1)}\bar{y}_M y_\pm{}^{-5}\bar{y}_S{}^{-2} \qquad \text{(XII-2)}$$

Since trace concentrations of M are used, $Q_{(1)} = D/[A^-]^4[\bar{S}]^2$ depends strongly on the concentration of nitric acid through the activity coefficient term for the aqueous phase, and one way to overcome this difficulty is to extrapolate to zero nitric acid concentration:

$$\log Q^{\infty}_{(1)} = \lim\,(C_{HA} \to 0)Q_{(1)} \qquad \text{(XII-3)}$$

Because of the hydrolysis of M^{4+}, reliable data cannot be obtained at

$C_{HA} < 0.1$ M, where the curve of D against C_{HA} is quite steep, so that the extrapolation is beset with some uncertainty. An illustration of this difficulty is seen by the formula given in (680):

$$Q_{(1)} = -0.43 + 3.11\mu^{-1} + 1.79\mu^{-2} \qquad \text{(XII-4)}$$

which cannot, of course, be extrapolated to $\mu = 0$, and is intended for use only down to $\mu = 0.6$. At any given HA concentration, the activity coefficient may be estimated from

$$y_{\pm MA_4}(\text{trace}) = (Q_{(1)}/Q^{\infty}_{(1)})^{1/5} \qquad \text{(XII-5)}$$

but this, of course, is not an independent estimate, contrary to what is implied by several workers (63P, 65S). No independent determination of $y_{\pm MA_4}$ has been made: the reference in (65S) to isopiestically determined values in (62R) is a mistake, since in that work no such data have been reported. However, independent estimates of the activity coefficient have been made. Hesford et al (58H) fitted $y^{1/4}_{\pm MA_4}$ into the family of curves of 1:1 electrolytes (cf. also the comparison of $y_{\pm MA_4}$ from (XII-4) with the experimental value for M = Th in (62R)), but took in addition into account the complexing of M with A, known independently. This is in principle a correct procedure, since $y_{\pm MA_4}$ is a stoichiometric activity coefficient, while the value estimated from the family of curves is a hypothetical one for completely dissociated $M^{4+} + 4A^{-}$, and must be corrected for complexing. However, the uncertainties of the fitting, together with that of the complexing, preclude the obtaining of an accurate value, so that

$$Q'_{(1)} = Q_{(1)}/y_{\pm MA_4}^{\ 5} \qquad \text{(XII-6)}$$

calculated in this way ($\log Q'_{(1)} = 5.20$) is quite remote from the value of $Q^{\infty}_{(1)}$ ($\log Q^{\infty} = 3$ to 3.4) which should be equivalent. Another attempt (64S) employed the parameter ρ_{Pu}, defined as the reciprocal of "the equivalent surface density of water molecules in the first coordination sphere of the ion" but did not report its value, or how it was obtained independently of the distribution data, to calculate an equilibrium constant,

$$\log C_p = 7.78 \pm 0.09 \qquad \text{(XII-7)}$$

on the mole fraction scale, which however cannot be compared with other reported constants for the reason stated. Finally, an attempt to estimate $y_{\pm MA_4}$ for trace plutonium in nitric acid by the use of Harned's rule (66E) seems to have been unsuccessful, since the resulting equilibrium constants (calculated from data reported in (60S) and unpublished sources) are unreasonably low ($\log Q'_{(1)}$ ranging from -0.3 to 1.0). Early estimates of the equilibrium constant gave only $Q_{(1)}$.* Subsequent work refers to some early publications, where data for the distribution are reported, but no equilibrium constants were calculated (57B, 58B, 58C, 60P, 60S, 60T), and sometimes also to irrevalent publications ((61M) is quoted in (64R) for data on extraction with TBP, whereas (61M) deals exclusively with TO'PO). The studies which report equilibrium constants deal either with undiluted S, or with its solutions in carbon tetrachloride or in kerosene. For neat S, (64Z) reports $Q_{(1)}$ at nine concentrations of HA from 0.1 to 5.5 M at 20±2°, extrapolating to $\log Q^{\infty}_{(1)} = 3.30$, and (62R) reports similarly $Q_{(1)}$ at five concentrations of HA from 0.1 to 2.0 M

---

*E.g., in (56F) "if activity coefficients are neglected" $\log Q_{(1)} = 0.48$ "fits most of the data for Pu(IV)," or, if $\log \beta_4 = -1.0$ is accepted for the formation of the aqueous species MA_4, $\log Q_{(8)} = 1.48$ is obtained (56Fa) from the above data for the reaction formulated as

$$MA_4 + 2\overline{S} = \overline{MA_4S_2} \qquad \text{(XII-8)}$$

at 20, 40 and 70°, extrapolating to log $Q^{\infty}_{(1)}$ = 3.10 at 20°. Finally, (65S) gives at 20° for 3.67 M S in CCl_4 -- which evidently refers to neat S, although the hydrated, undiluted S is only 3.43 M (Section V) -- the value

$$\log Q'_{(1)} = 3.05 \pm 0.07, \qquad \text{(XII-9)}$$

where the "dispersion of the data is determined ... by random errors." The first report gives the constant to only one significant figure, while the second one gives an unreasonably low value for log $Q^{\infty}_{(1)}$ at 70° (2.38), although the value at 40° is reasonable (3.00), which may reflect some difficulty at the elevated temperature, or a more general inaccuracy. Thus, the best value that can be given is

$$\log Q'_{(1)} = \log Q^{\infty}_{(1)} = 3.07 \pm 0.07 \qquad \text{(XII-10)}$$

for undiluted S at 20°.

According to discussions in a previous section, the standard state for the organic phase is hydrated, undiluted S, so that for the conditions for which $Q^{\infty}_{(1)}$ has been determined

$$\bar{y}_M = \bar{y}_S = 1 \quad \text{and} \quad Q^{\infty}_{(1)} = K^{o}_{(1)}. \qquad \text{(XII-11)}$$

For carbon tetrachloride solutions, data have been given for 0.5 M S and varying HA concentrations: (63P) employing data of (58B, 60P and 60T), and (64R) employing untraceable data since the reference (61M) was misquoted, and for nominally 0.37 and 1.1 M (i.e. 10 and 30% vol. S) in CCl_4, again for varying HA (65S). Activity coefficients were estimated by (XII-4) from distribution data in kerosene given by (62R), and $Q'_{(1)}$ values were then calculated, and found independent of HA concentration below 4 M, and of S concentration over the whole range (65S). (Compare the similar behavior in CCl_4 of the uranium (VI) system described in Section XI-C.) The data of (60P) and (65S) refer to 20°, for the others the temperature was not specified. The constant log $Q'_{(1)}$ = 3.55 calculated (63P) from the data of (60P) and (60T) seems to be too high, if the uranium (VI) system (Section XII) is taken as a guide for the solvent effect, and the values log $Q'_{(1)}$ = 3.06 calculated (63P) from the data of (58B), log $Q'_{(1)}$ = 3.13 of (64R) and log $Q'_{(1)}$ = 3.07±0.02 of (65S) are more consistent. The last mentioned value is most likely to be the best estimate.

For kerosene solutions (hydrogenated kerosene (60M), kerosene of boiling range 140-240° (59R) or 188-222° (60S), or Gulf-BT or Amsco 123-15 (58C) are some specifications) data were given mostly at the single concentration of 20% vol. or 0.72 M (rounded to 0.7 M in some papers), with a value of log $Q^{\infty}_{(1)}$ = 3.42 (60M, 62R) at 20° and 3.40 (59R) at 25°. The data presented in (60M) are the most detailed, and their internal precision is ±10%, leading to an estimated error in log $Q^{\infty}_{(1)}$ of ±0.05. That the value is higher than for neat S or for CCl_4 solutions is consistent with a similar behavior of the uranium (VI) system (Section XI), and is evidently due to a similar behavior of $\overline{y}_M$ in the two systems. However, absence of data at lower S concentrations precludes the calculation of $Q'_{(1)}$ for the standard state of infinite dilution of S in the diluent. If the uranium (VI) system is again taken as a guide, a value of

$$\log \lim Q^{\infty}_{(1)} \ (\overline{C}_S \rightarrow 0) \sim 3.7 \qquad \text{(XII-12)}$$

can be estimated.

Table XII-1 summarizes the values selected as most reliable.

Table XII-1

Equilibrium constants for the reaction $Pu^{4+} + 4NO_3^- + 2\overline{TBP} = \overline{Pu(NO_3)_4(TBP)_2}$ at 20°

Diluent	log $Q^{\infty}_{(1)}$ at 0.7 M S	lim log $Q^{\infty}_{(1)}$ ($C_S \to 0$)	log $K^{o}_{(1)}$
Carbon tetrachloride	3.07±0.02	3.1	
Kerosene	3.42±0.05	3.7	
Undiluted			3.07±0.07

References:

56F Flanary, J.R., in "Progress in Nuclear Energy, Series III, Process Chemistry," Pergamon, Vol. 1, 195 (1956).

56Fa Fletcher, J.M., in "Progress in Nuclear Energy, Series III, Process Chemistry," Pergamon, Vol. 1, 105 (1956).

57B Best, G., McKay, H.A.C. and Woodgate, P., J. Inorg. Nucl. Chem., 4, 315 (1957).

58B Burger, L.L., J. Phys. Chem., 62, 590 (1958).

58C Codding, J.W., Haas, W.O. and Heumann, F.K., Ind. Eng. Chem., 50, 145 (1958).

58H Hesford, E. and McKay, H.A.C., Trans. Faraday Soc., 54, 573 (1958).

59R Rozen, A.M. and Moiseenko, E.I., Zh. Neorg. Khim., 4, 1209 (1959).

60M Moiseenko, E.I. and Rozen, A.M., Radiokhimiya, 2, 274 (1960).

60P Petrov, K.A., Shevchenko, V.B., Timoshev, V.G. et al., Zh. Neorg. Khim., 5, 498 (1960).

60S Shevchenko, V.B., and Fedorov, I.A., Radiokhimiya, 2, 6 (1960).

60T Timoshev, V.G., Petrov, K.A., Rodionov, A.V. et al., Radiokhimiya, 2, 419 (1960).

61M Martin, B., Ockenden, O.W. and Foreman, J.K., J. Inorg. Nucl. Chem., 21 96 (1961).

62R Rozen, A.M. and Moiseenko, E.I., in "Extraktsiya," Atomizdat, Vol. 2, 236 (1962).

63P Pushlenkov, M.F. and Shuvalov, O.N., Radiokhimiya, 5, 543 (1963).

64R Rozen, A.M. and Nikolotova, Z.I., Zh. Neorg. Khim., 9, 1725 (1964).

64S Solovkin, A.S., Zh. Neorg. Khim., 9, 746 (1964).

64Z Zemlyanukhin, V.I., Savoskina, G.P. and Pushlenkov, M.F., Radiokhimiya, 6, 714 (1964).

65S Shuvalov, O.N., Radiokhimiya, 7, 265 (1965).

66E Estrada, J., Diaz-Nogueira, E. and Lopez-Perez, B., Ann. Real Soc. Espana Fis. Quim. Ser. B62, 231 (3) (1966).

67O Ochsenfeld, W., Schmieder, H., Baumgärtel, G. and Kuhn, E., German Report KFK-671 (1967).

XIII. Solvent: Tri-n-octyl phosphine oxide (TO'PO) , S

Ligand: Nitrate (NO_3^-), A^-

Reaction: $H^+ + A^- + \overline{S} = \overline{HAS}$ (XIII-1)

TO'PO is a rather reactive reagent, with a basicity much exceeding that of TBP. Hence, its interactions with water, acids or metal salts are expected to be relatively strong. Its limited solubility in inert solvents (at 25° a saturated solution in cyclohexane is 0.992 M, with $\overline{x}_S = 0.147$ (60Z)) detracts somewhat from its usefulness. Its solubility in water, however, is so much lower, i.e. for

$$S = \overline{S} \qquad \text{(XIII-2)}$$

$\log Q_{(2)} > 5$ for 0.1 M S in benzene solution (67G), that no losses to the aqueous phase, or effects of the reagent on equilibria and activities in the aqueous phase, need be feared.

TO'PO does, however, react strongly with water, forming the monohydrate:

$$H_2O + \overline{S} = \overline{H_2O \cdot S} \qquad \text{(XIII-3)}$$

with an equilibrium quotient $Q_{(3)}$ (on a mixed scale, with $a_{H_2O} = 1$), obtained after correction for water extracted by the diluent alone. The values

$$\log Q_{(3)} = -0.25 \pm 0.03 \qquad \text{(XIII-4)}$$

for carbon tetrachloride solutions (65C, 670) and

$$\log Q_{(3)} = 0.02 \pm 0.08 \qquad \text{(XIII-5)}$$

for nitrobenzene solutions (65C) are consistent with the data. Low values for the saturation water concentrations of dilute (0.1 - 0.2) solutions of S in several diluents (O-dichlorobenzene, carbon tetrachloride and cyclohexane), in the range 0.43 to 0.62 H_2O : S (65M) are inconsistent with the above data, and with the expected behavior of S, relative to the less basic TBP (Section VI), and must be discounted.

An attempt to obtain the activity coefficients of S in n-octane (62B) was not quite successful, since the isopiestic standard employed, triphenylmethane was apparently non-ideal in the diluent, but its own activity coefficient could not be established accurately. Furthermore, doubt exists regarding whether activities established in dry S-octane solutions are valid also for a wet organic phase.

The stoichiometry of reaction (XIII-1) has been established with reasonable certainty in both saturation (60Z, 61M) and distribution studies. This equilibrium predominates by far at aqueous HA concentrations $\leq$ 2M. At higher acidities excess HA is extracted, and it is unclear whether associated HA is extracted into the polar diluent - HAS mixture, or whether definite new species are formed.

Equilibrium (XIII-1) seems to be not very strongly dependent on the diluents, since a variety of them seems to yield a value of log $Q_{(1)}$ ranging from 0.96 to 1.35. However, two situations must be distinguished. In one, a constant ionic medium is used, provided in all cases by sodium nitrate, NaA, of 2.0 M (60Z, 67G) or 0.40 M (64H). The equilibrium quotient

$$Q_{(1)} = [\overline{HAS}]/[H^+][A^-][\overline{S}] \qquad \text{(XIII-6)}$$

was found to be independent of C_{HA} and of $\overline{C}_S$, the former at low concentrations compared to the ionic medium, where $y_{\pm HA}$ is constant, the latter up to 0.2 M (depending on the diluent). The constancy of $Q_{(1)}$ is taken to indicate that $\overline{y}_{HAS}/\overline{y}_S$ is independent of $\overline{C}_{HA}$ and $\overline{C}_S$ at low concentrations. The early value obtained for cyclohexane and 2 M NaA solutions, log $Q_{(1)}$ = 0.96 (60Z) seems to be somewhat low. A later value, obtained however for 0.4 M NaA,

$$\log Q_{(1)} = 0.99 \pm 0.03 \qquad \text{(XIII-7)}$$

(64H) would correspond to a higher value at 2 M NaA, because of salting out; however, no data are available on $y_{\pm HA}$ in NaA for applying the correction. The values log $Q_{(1)}$ = 1.35±0.02 for n-octane and 1.18±0.03 for benzene and 2.0 M NaA medium (67G) have been well documented and can be accepted as valid.

In another series of reports, varying HA concentrations have been used with no extraneous nitrate, so that

$$Q'_{(1)} = [\overline{HAS}]/a_{\pm HNO_3}[\overline{S}] = [\overline{HAS}]/[H^+]^2 y_{\pm HA}{}^2[\overline{S}] \qquad \text{(XIII-8)}$$

can be evaluated. The data of (63M) obtained for 1 M HA and varying S concentrations, can be converted to

$$\log Q'_{(1)} = \log Q_{(1)} - 2 \log y_{\pm HA} = 1.01\pm0.07 - 2(-0.125) = 1.26\pm0.07 \quad \text{(XIII-9)}$$

for dichlorobenzene. The data of (64R) include an extrapolation to

$$\lim Q_{(1)} \ (\overline{C}_{HA} \rightarrow 0) = 1.15\pm0.02 \quad \text{(XIII-10)}$$

(from four points only) and since at the limit for $\overline{C}_{HA}$ also $C_{HA} \rightarrow 0$ and $y_{\pm HA} \rightarrow 1$, the above value represents $\log Q'_{(1)}$ for carbon tetrachloride, in good agreement with another reported value for this solvent, based on many more data,

$$\log Q'_{(1)} = 1.12\pm0.02 \quad \text{(XIII-11)}$$

(63C). Finally, for benzene, the value

$$\log Q'_{(1)} = 1.18\pm0.03 \quad \text{(XIII-12)}$$

(67G) has been obtained, again well documented. All these values were obtained at 25°C. They indicate that $\overline{y}_{HAS}/\overline{y}_S$ is constant within the concentration range employed, up to 0.1 M S in the diluent and up to $\overline{C}_{HA} \sim \overline{C}_S$. The results can be extrapolated to $\overline{C}_S \rightarrow 0$, hence also $[\overline{HAS}] \rightarrow 0$, so that $Q'_{(1)} = K^{\infty}_{(1)}$ is a good approximation, $K^{\infty}_{(1)}$ pertaining to the standard states of infinite dilution of all solutes in water and in the (hydrated) diluent. The results are summarized in Table XIII-1.

Table XIII-1

Equilibrium quotients for the reaction $H^+ + NO_3^- + \overline{TO'PO} = \overline{HNO_3 \cdot TO'PO}$ at 25°

Diluent	$\log Q'_{(1)} \approx \log K_{(1)}$	Ref.
Benzene	1.18±0.03	67G
o-Dichlorobenzene	1.26±0.07	63M
Carbon tetrachloride	1.13±0.02	63C, 64R

References:

60Z Zingaro, R.A. and White, J.C., J. Inorg. Nucl. Chem., 12, 315 (1960).

61M Martin, B., Ockenden, D.W. and Foreman, J.K., J. Inorg. Nucl. Chem., 21, 96 (1961).

62B Baes, C.F., J. Phys. Chem., 66, 1629 (1962).

63C Conocchioli, T.S., M.Sc. Thesis, Univ. California, Berkeley, 1963, Report UCRL - 10971 (1963).

63M Mrochek, J.E., O'Laughlin, J.W., Sakurai, H. and Banks, C.V., J. Inorg. Nucl. Chem., 25, 955 (1963).

64H Heyn, A.H.A. and Soman, Y.D., J. Inorg. Nucl. Chem., 26, 287 (1964).

64R Rozen, A.M. and Nikolotova, Z.I., Zh. Neorg. Khim., 9, 1725 (1964).

65C Conocchioli, T.S., Tocher, M.I. and Diamond, R.M., J. Phys. Chem., 69, 1106 (1965).

65M Mrochek, J.E., Richard, J.J. and Banks, C.V., J. Inorg. Nucl. Chem., 27, 625 (1965).

67G Goffart, J. and Duyckaerts, G., Anal. Chim. Acta, 38, 529 (1967).

67O O'Laughlin, J.W. and Banks, C.V. in "Solvent Extraction Chemistry," (D. Dyrssen, J.-O. Liljenzin and J. Rydberg, Eds.), North-Holland, 1967, p. 273.

XIV. Solvent: Tri-n-butyl phosphine oxide (TB'PO) , S

Ligand: Nitrate (NO_3^-), A^-; Metal: uranium (VI) (UO_2^{2+}), M^{2+}

Reaction: $$M^{2+} + 2A^- + 2\overline{S} = \overline{MA_2S_2} \quad \text{(XIV-1)}$$

It was realized a considerable time ago that exchanging an alkoxy group in phosphate esters for an alkyl group strongly enhances the extracting ability for M. It was thus natural to examine the effect of exchanging all three groups in TBP, yielding the phosphine oxide S. However, none of the studies concentrated on this particular solvent, and its extractive power has always been examined, relative to other extractants, in a rather superficial manner only. Thus, some authors (63P, 64R) calculated the "effective equilibrium constant" for reaction (XIV-1),

$$Q'_{(1)} = D_M/4(A^-)^2 y_{\pm MA_2}{}^3 (\overline{S})^2 \quad \text{(XIV-2)}$$

from only a single value of D_M, given for one particular concentration of each reactant (58B, 59H). Other authors who calculated $Q'_{(1)}$ also did not document their calculated values with extensive data, so that the apparent quality of the constants reported must remain rather poor.

For the diluent carbon tetrachloride, the single datum of (58B), obtained for initial concentrations $\overline{C}_S$ = 0.50 M and C_M = 0.20 M with D_M = 23 leads to

$$\log Q = \log \{[\overline{M}]/4[M]^3(\overline{C}_S - 2[\overline{M}])^2\} = 6.79 \quad \text{(XIV-3)}$$

where $[\overline{M}] = C_M D_M/(1+D_M)$ and $[M] = C_M/(1+D_M)$ (64R). Various estimates of $y_{\pm MA_2}$ at the equilibrium aqueous concentration [M] = 8.33 10^{-3}M then led to log Q' = 7.13 (63P) or 7.20 (64R). Another study of MA_2 solutions without excess HA suffered from a "considerable dispersion on account of the complexity of determining small values of $[\overline{M}]$ and $\overline{C}_S$" (63P), leading to a very inaccurate value of log $Q'_{(1)}$ = 7.0±0.3.

Several other studies employing carbon tetrachloride as a diluent deal with dilute solutions of MA_2 in HA. An early study (59V, cf. also for same data 60P) utilized mixtures of HA and $HClO_4$ and gives three different values of log $Q_{(1)}$: 6.40 for a single point, referring to a solution of 0.714 M HA; 6.54 as one parameter (referring to pure A^- solutions) out of three parameters of an equation for mixed HA-$HClO_4$ solutions (for 7 different compositions); 6.58 as the selected value for log $Q_{(1)}$, with no justification of the choice. In all the solutions, μ = 0.714 M, $C_M = 2.9 \times 10^{-5}$ M and $\overline{C}_S = 8.4 \times 10^{-3}$ M were kept constant. Application of the ionic strength principle (see Section XII) leads to log $y_{\pm MA_2}$ (for C_M = 0.24 M = $\mu/3$) = -0.279, and to log $Q'_{(1)}$ = 7.24, 7.38 and 7.42 respectively. The first value was selected (63P) to represent the results of this study, in fair agreement with the estimates quoted above for log $Q'_{(1)}$.

In spite of this agreement, these values of log $Q'_{(1)} > 7$ could be considerably too high, in view of further data that have been published for carbon tetrachloride-diluent systems on the one hand, and to benzene-diluent systems on the other. A report (60Pa) giving three points, at 0.5, 1.0 and 2.0 M HA leads to log $Q_{(1)}$ = 6.08, 5.33 and 3.95 respectively, from which the value log $Q'_{(1)}$ = 6.50 has been calculated (63P) with no details. This refers to the total extractant concentration $\overline{C}_S$ = 0.50 M. At another appreciable concentration of extractant, 0.73 M and higher HA concentrations (1-10 M), extrapolation to $\sqrt{\mu} = 0$, which should give a $Q'_{(1)}$ value, yielded log $Q'_{(1)}$ = 6.40 (61N), but a double extrapolation to $C_{HA} \rightarrow 0$ and $\overline{C}_S \rightarrow 0$, which should yield $Q'^{\infty}_{(1)}$, yielded the rather low value log $Q'_{(1)}$ = 5.71 (63A). Since no information on the activity coefficients of the MA_2S_2 is available (S in CCl_4 has been studied, (63P)), there is no way to check the reason for this low value.

In benzene solutions, early work (59H) gave a single datum at $\overline{C}_S$ = 0.19 M,

C_{HA} = 0.5 M and trace MA_2. If the equation

$$[\bar{S}] = \bar{C}_S/(1 + 8.4\ C_{HA}^2) \qquad \text{(XIV-4)}$$

is used to correct for S bound to HA, implying the binding of two HA per S, for which (64R), there is no evidence, log $Q_{(1)}$ = 5.76 results. Alternatively, log Q = 6.39 if $[\bar{S}] = \bar{C}_S/(1+8.4C_{HA})$ is used instead, accepting for the sake of argument the equilibrium quotient 8.4 for binding the HA. Application of the "ionic strength principle" then leads to log $Q'_{(1)}$ = 6.61 (64R) or 7.20, if the more reasonable binding of a single HA is accepted for the correction term. Another (63P) estimate of log $Q'_{(1)}$ = 5.92 from the same datum (59H), but ignoring the binding of HA, is evidently too low. In another study, (62S), giving five data points at $\bar{C}_S = 3.54\times10^{-4}$ M and 2 M HA yields again a rather low value

$$\log Q_{(1)} = 4.63\pm0.05 \qquad \text{(XIV-5)}$$

since the binding of a large fraction of S by the coextracted HA was ignored. Applying the above correction (single HA bound) to convert from $\bar{C}_S$ to [S] and the "ionic strength principle" for calculating $y_{\pm MA_2}$ yields log $Q'_{(1)}$ = 7.94. This is possibly somewhat too high, but on comparison with TBP (Section XII), it can be expected that $Q'_{(1)}$ should be somewhat higher for benzene than for carbon tetrachloride.

The data shown (63P) for chloroform solutions (four datum points) lead to a value of log $Q'_{(1)}$ = 4.75, compared with 7.0 for carbon tetrachloride. This is a reasonable value for this associating diluent.

In view of the inconsistencies in the $Q'_{(1)}$ values quoted, it does not matter whether the values of $Q_{(1)}$ are for 20° (60Pa), 22° (59H), room temperature (62S), or 25° (58B, 59V, 61N), since the inaccuracies discussed above far outweigh the effects of inaccuracies in the temperature. The values reported in Table XIV-1 must be considered as very tentative, and must be checked with additional data.

Table XIV-1

Equilibrium quotients for $UO_2^{2+} + 2NO_3^- + 2\overline{TB'PO} = \overline{UO_2(NO_3)_2(TB'PO)_2}$ at 20-25°C

Diluent	CCl_4	C_6H_6	$CHCl_3$
log $Q'_{(1)}$	7.2 (?)	7.2 to 7.9 (?)	4.8 (?)

References:

58B Burger, L.L., J. Phys. Chem., 62, 590 (1958).

59H Healy, T.V. and Kennedy, J., J. Inorg. Nucl. Chem., 10, 128 (1959).

59V Voden, V.G., Nikitina, G.P. and Pushlenkov, M.P., Radiokhimiya, 1, 121 (1959).

60P Pushlenkov, M.P., Nikitina, G.P. and Voden, V.G., Radiokhimiya, 2, 215 (1960).

60Pa Petrov, K.A., Shevchenko, V.B., Timoshev, V.G. et al., Zh. Neorg. Khim., 5, 498 (1960).

61N Nomura, S. and Hara, R., Anal. Chim. Acta, 25, 212 (1961).

62S Shevchenko, V.B., Smelov, V.S. and Strakhova, A.V., Zh. Neorg. Khim., 7, 1736 (1962).

63N Nishi, T. and Asano, M., J. Chem. Soc. Japan, Ind. Chem. Sec., 66, 1428 (1963).

63P Pushlenkov, M.P. and Shuvalov, E.V., Radiokhimiya, 5, 551 (1963).

64R Rozen, A.M. and Nikolotova, Z.I., Zh. Neorg. Khim., 9, 1725 (1964).

XV. Solvent: Di-n-butyl n-butylphosphonate (DBB'P), S

Ligand: Nitrate (NO_3^-), A^-

Reaction: $H^+ + A^- + \overline{S} = \overline{HAS}$ (XV-1)

In systematic studies of the effect of exchanging alkoxy for alkyl groups in phosphate esters on their extraction abilities, it was found that S is a convenient extractant. Several studies include its reaction with HA, showing it to proceed according to (XV-1) (59V, 63P). The equilibrium quotient

$$Q_{(1)} = [\overline{HAS}]/[H^+][A^-][\overline{S}] \qquad \text{(XV-2)}$$

and the corrected quotient

$$Q'_{(1)} = Q_{(1)}/y_{\pm HA}^2 \qquad \text{(XV-3)}$$

have been reported for a number of diluents, and for neat S (that is, hydrated, undiluted S). An estimate of

$$\log \overline{\gamma}_{HAS} = 0.29\ \overline{m}_{HAS} \qquad \text{(XV-4)}$$

in n-hexane has been made (59B), and estimates of $\overline{f}_S$ for n-hexane and carbon tetrachloride can be made from available data of $\overline{f}_{diluent}$ (63P), through the Gibbs-Duhem relationship. In those cases where $Q'_{(1)}$ is found to be independent of $\overline{C}_S$ down to low values of this variable, the ratio $\overline{y}_{HAS}/\overline{y}_S$ must be constant to the limit of infinite dilution,

$$\lim Q'_{(1)}\ (\overline{C}_S \rightarrow 0) = Q'^{\infty}_{(1)} = K^{\infty}_{(1)} \qquad \text{(XV-5)}$$

where each activity coefficient equals unity, so that the equilibrium quotient equals the thermodynamic equilibrium constant.

For carbon tetrachloride solutions, at a given HA concentration, 1.16 M, and varying $\overline{C}_S$, from 0.03 to 0.18 M, 6 experimental points lead to

$$\log Q'_{(1)} = -0.64 \pm 0.04. \qquad \text{(XV-6)}$$

This value is somewhat low, compared with the finding, undocumented with data (63P), that

$$\log Q'_{(1)} = \log Q'_{(1,\ \text{for S=TBP})} + 0.42 \qquad \text{(XV-7)}$$

which from Table VII-1 comes out to be -0.42 ± 0.13. This value agrees with the value $\log Q'_{(1)} = -0.41$ calculated from the Gibbs energy change for 0.15 M solutions of S (63N).

For hydrocarbon diluents, the scattering of the values reported is larger. From a single point at 0.05 M S in n-dodecane and 1 M HA, the value

$$\log Q'_{(1)} = -0.63 \pm 0.01 - 2(-0.14) = -0.35 \pm 0.01 \qquad \text{(XV-8)}$$

(59S) is obtained, in reasonable agreement with $\log Q'_{(1)} = -0.29$ (63N), but about a factor of two lower than $\log Q'_{(1)} = 0.00$ for 10% vol. (i.e. ca. 0.35 M) S in kerosene (64Z), or

$$\log Q'_{(1)} = \log Q'_{(1,\ \text{S=TBP})} + 0.5 \qquad \text{(XV-9)}$$

(63P), which from Table VII-1 yields $+0.08 \pm 0.09$ for kerosene. A surprisingly high value,

$$\log Q'_{(1)} = \log Q'_{(1)}(\text{m-scale}) + \log d_{\text{hexane}} = +0.37 \qquad \text{(XV-10)}$$

has been obtained from an equation that includes a correction for $\bar{\gamma}_{HAS}$ and a_{H_2O} (59B). This equation, which was not supported by reported data, requires $Q'_{(1)}$ to vary with $\bar{m}_{HAS}$, hence also with C_{HA}, and this is at variance with the findings of other authors, and cannot be accepted without further evidence. Xylene shows a value of $\log Q_{(1)} = -0.3$, or $\log Q'_{(1)} = 0.0$ (64S) in line with the aliphatic hydrocarbons for 0.7-1.1 M S and varying C_{HA}. The expected salting-out by NaA has also been observed.

Finally, for neat S, the value $\log Q'^{\infty}_{(1)} = +0.65$ has been obtained (64Z, 64Za), on extrapolation of 11 data points from $C_{HA} = 2.7$ M down to $C_{HA} = 0.12$ M and further to $C_{HA} = 0$. A comparison with the corresponding data for S = TBP (Table VII-1) shows this value to be reasonable, since

$$\log Q' \text{ (undiluted)} - \log Q' \text{ (kerosene)} \simeq 0.4 \pm 0.2 \qquad \text{(XV-11)}$$

in that case.

It is thus seen that different diluents do not lead to widely varying values of $\log Q'_{(1)}$, but that there is a definite negative Gibbs energy of transfer from trace HA in neat S to trace HA in infinitely dilute S. The results are summarized in Table XV-1.

Table XV-1

Equilibrium constants for $H^+ + NO_3^- + \overline{DBB'P} = \overline{HNO_3 \cdot DBB'P}$ at 20-25°

Diluent	CCl_4	Hydrocarbons	Undiluted
$\log Q'_{(1)}$	-0.4±0.1	-0.1±0.1	+0.6±0.1

References:

59B Baldwin, W.H., Higgins, C.E. and Soldano, B.A., J. Phys. Chem., 63, 118 (1959).

59S Siddall, T.H., J. Am. Chem. Soc., 81, 4176 (1959).

59V Voden, V.G., Nikitina, G.P., and Pushlenkov, M.F., Radiokhimiya, 1, 121 (1959).

63N Nishi, T. and Asano, M., J. Chem. Soc. Japan, Ind. Chem. Sec., 66, 1428 (1963).

63P Pushlenkov, M.F. and Shuvalov, O.N., Radiokhimiya, 5, 551 (1963).

64S Sheppard, J.S., USAEC Report HW-81166 (1964).

64Z Zemlyanukhin, V.I., Savoskina, G.P. and Pushlenkov, M.F., Radiokhimiya, 6, 694 (1964).

64Za Zemlyanukhin, V.I., Savoskina, G.P. and Pushlenkov, M.F., Radiokhimiya, 6, 714 (1964).

XVI. Solvent: Di-n-butyl n-butylphosphonate (DBB'P) , S

Ligand: Nitrate (NO_3^-), A^-; Metal: Uranium (VI) (UO_2^{2+}), M^{2+}

Reaction: $M^{2+} + 2A^- + 2\overline{S} = \overline{MA_2S_2}$ (XVI-1)

As a member of the series $(C_4H_9O)_n(C_4H_9)_{3-n}PO$, in which the exchange of alkyl for the alkoxy groups of TBP was shown to enhance the extraction of MA_2 very considerably, the solvent S has received its share of attention. The equilibrium quotient for (XVI-1), corrected for the activity coefficient of MA_2 in the aqueous phase is

$$Q'_{(1)} = [\overline{MA_2S_2}]/[M^{2+}][A^-]^2 y_{\pm MA_2}{}^3 [S]^2 \quad \text{(XVI-2)}$$

As before, in those cases where trace MA_2 in solutions of HA was used, $y_{\pm MA_2}$ was estimated from the "ionic strength principle" (Section XI), so that for < 2 M HA, 3 log $y_\pm$ = 0.81±0.09. If $Q'_{(1)}$ is found to be effectively constant over a concentration range of the reactants, it follows that the ratio $\overline{y}_{MA_2S_2}/\overline{y}_S{}^2$ is constant over this range.

For hydrocarbon diluents, (59S) gives a highly precise value of log $Q_{(1)}$= 2.94±0.03 for 25°C, based on 26 experimental points at temperatures varying from 0° to 50°, but at only one HA concentration (with trace M), 1.0 M. It was argued that 0.05 M is sufficiently low for the organic phase solutions to be ideal.

The value

$$\log Q'_{(1)} = 3.75 \pm 0.09 \qquad \text{(XVI-3)}$$

is obtained from this on correction for $y_{\pm MA_2}$. The values of log $Q'_{(1)}$ calculated from the single datum point at 0.19 M S in benzene and 0.5 M HA (59H): 3.94 (64R) or 3.89 (63P) are consistent with the above estimate, accepting the similarity of the values of Q' for the analogous reaction for S = TBP (Table XI-1). The value log $Q_{(1)}$ = 3.40 (63N), on the other hand, is definitely too high, since it leads to log $Q'_{(1)}$ = 4.21.

For carbon tetrachloride solutions, there are data for 0.03-0.18 M S at 1.16 M HA and trace M (6 experimental points (59V)), which give log $Q_{(1)}$ = 2.68±0.03 and log $Q'_{(1)}$ = 3.49±0.09. For undisclosed reasons, the authors (59V) prefer the higher value log $Q_{(1)}$ = 2.98, obtained in mixed HA-$HClO_4$ solutions at μ = 0.714 extrapolated to C_{HA} = 0, but this leads to unknown activity coefficients (of MA_2 in $HClO_4$ solutions), so that $Q'_{(1)}$ cannot be calculated. Later work by the senior author (60P) gives a lower value of

$$\lim Q_{(1)}(C_{HA} \rightarrow 0 \text{ at } \mu = 0.714) = 2.78 \qquad \text{(XVI-4)}$$

but again this cannot be converted to $Q'_{(1)}$, while still later work (63P) reports the results of calculations from (59V, 60P) as yielding log $Q'_{(1)}$ = 3.46, with no details of the calculation. This last mentioned work gives log $Q'_{(1)}$ = 3.41±0.06 from 9 data points at [S]= $(0.2\text{-}5.0 \times 10^{-3}$ M and $C_{M(total)}$ = 0.18 M. Other, less precise, values have been calculated from data in (58B) log $Q'_{(1)}$ = 3.25 by (63P) and 3.23 by (64R) for two data points at $\overline{C}_S$ = 0.5 M and $C_{M(total)}$ = 0.2 and 0.05 M, and from data in (60Pa, 60T) by (63P), log $Q'_{(1)}$ = 3.62 for $\overline{C}_S$ = 0.5 M, trace M and 0.5, 1.0 and 2.0 M HA. The difference

$$\log Q_{(1)}(\text{hydrocarbon diluent}) - \log Q_{(1)}(CCl_4 \text{ diluent}) = 0.27 \qquad \text{(XVI-5)}$$

(63N) is consistent with the best values obtained for the individual diluents (Table XVI-1).

Finally, an estimate of log $Q^{\infty}_{(1)}$ = 2.27 for lim $Q_{(1)}(C_{HA} \rightarrow 0)$ for neat, undiluted but hydrated S, has been made (64Za) for trace M in 0.12-2.7 M HA. This value seems to be rather low, compared to those obtained in the diluents, in view of the data in Table XI-1 (log $Q'_{(1)(undiluted)}$ - log $Q'_{(1)(hydrocarbon)}$ = -0.5) for S = TBP. The value for neat S cannot, therefore, be considered as established.

Table XVI-1

Equilibrium constants for $UO_2^{2+} + 2NO_3^- + 2\overline{DBB'P} = \overline{UO_2(NO_3)_2(DBB'P)_2}$ at 20-25°

Diluent	Hydrocarbon	CCl_4
log $Q'_{(1)}$	3.75±0.09	3.46±0.05

References:

58B Burger, L.L., J. Phys. Chem., 62, 590 (1958).

59H Healy, T.V. and Kennedy, J., J. Inorg. Nucl. Chem., 10, 128 (1959).

59S Siddall, T.H., J. Am. Chem. Soc., 81, 4176 (1959).

59V Voden, V.G., Nikitina, G.P. and Pushlenkov, M.F., Radiokhimiya, 1, 121 (1959).

60P Pushlenkov, M.F., Nikitina, G.P. and Voden, V.G., Radiokhimiya, 2, 215 (1960).

60Pa Petrov, K.A., Shevchenko, V.B., Timoshev, V.G. et al., Zh. Neorg. Khim., 5, 498 (1960).

60T Timoshev, V.G., Petrov, K.A., Rodionov, A.A. et al., Radiokhimiya, 2, 419 (1960).

63N Nishi, T. and Asano, M., J. Chem. Soc. Japan, Ind. Chem. Sec., 66, 1428 (1963).

63P Pushlenkov, M.F. and Shuvalov, O.N., Radiokhimiya, 5, 551 (1963).

64R Rozen, A.M. and Nikolotova, Z.I., Zh. Neorg. Khim., 9, 1725 (1964).

64Z Zemlyanukhin, V.I., Savoskina, G.P. and Pushlenkov, M.F., Radiokhimiya, 6, 714 (1964).

Made in the USA
Monee, IL
13 September 2026

62978990R10057